A NATURAL HISTORY OF QUIET WATERS

A Natural History of Quiet Waters

Swamps and Wetlands of the Mid-Atlantic Coast

CURTIS J. BADGER

University of Virginia Press
CHARLOTTESVILLE AND LONDON

University of Virginia Press

Printed in the United States of America on acid-free paper

First published 2007

1 3 5 7 9 8 6 4 2

Library of Congress Cataloging-in-Publication Data
Badger, Curtis J.
A natural history of quiet waters : swamps and wetlands
of the mid-atlantic coast / Curtis J. Badger.
p. cm.
ISBN 978-0-8139-2618-6 (cloth : alk. paper)
1. Wetlands—Middle Atlantic States. 2. Natural history—
Middle Atlantic States. 3. Wetland ecology—Middle Atlantic States.
4. Swamps—Middle Atlantic States. I. Title.
QH87.3.B33 2007
578.7680975—dc22
2007011559

To George T. Parker,

a lifelong friend who introduced me
to the pleasures of paddling a canoe many years ago,
and who shares my appreciation for swamps
and the things that live there

Contents

Acknowledgments

MY WIFE, LYNN, and son, Tom, have shared with me many of the adventures described herein. Shortly after we got married, Lynn and I got lost in the dark in the woods along Pungoteague Creek after staying out too late on a sunset canoe trip. We eventually found our way home, our marriage survived, and Tom later came along to join us on many subsequent trips. We now carry a flashlight.

Many of the special places described here would not be so if it were not for the work of groups such as the Chesapeake Bay Foundation and The Nature Conservancy. I salute both of them, and I especially want to thank John Page Williams of CBF for discussing water quality with me, as well as Barry Truitt and Brian van Eerden of the conservancy. Barry introduced me to radar ornithology and Brian taught me a great deal about longleaf pines. Thanks also to Mike Watson, NASA radar expert and radar ornithologist.

John McCoy of the Maryland Department of Natural Resources discussed issues of poultry farming and water quality with me, and Kevin Smith showed me the restored wetlands of Marshyhope Creek. Hunting guide and decoy carver Grayson Chesser introduced me to Pitts Creek on the Eastern Shore of Virginia, a pristine wetland threatened by residential development.

The Nature Conservancy owns a preserve of more than 9,000 acres along the Nassawango Creek in Maryland. Joe Fehrer, whose father was instrumental in creating the preserve, introduced me to the plants and animals of this great place. Situated in the Nassawango Preserve is Furnace Town, a restored village whose mission in the early 1800s was to make iron from bog ore found in the Nassawango swamp. Kathy Fisher led me through the history of the village and its role in iron-making. An excellent naturalist, she also taught me much about plants of the Nassawango.

And special thanks to realtor Carole Gubb, who, as this book was being written, found us our own special place on Pungoteague Creek, a modest little parcel of freshwater wetland, salt marsh, and upland. We now call it home.

A NATURAL HISTORY OF QUIET WATERS

1

Real Estate of Ill Repute

AND SO I CALLED my friend Carole Gubb, who sells real estate. "Carole," I said, "I want to buy fifty acres of worthless swampland."

"Boy, have I got a deal for you," she said.

Only, she didn't.

I like swamps. I like the earthy way they smell after a rain. I like the dark, quiet water that flows almost imperceptably, strong and deep, flexing with the tide like steel. I like the mystery of a swamp, the uncertainty of what could be hiding beyond that fold of cypress and black gum. I enjoy the notorious reputation of swamps, their kinship to criminals and pirates and others of ill repute. Swamps are like graceful older women who were wild and wanton at nineteen and twenty, who experienced a life that makes them beautiful and desirable because of what they may have known.

I wanted a swamp of my own, and I had recently come into a little bit of money. Not any dazzling amount, mind you, no sell-the-house-buy-a-yacht-we're-leaving-for-Tahiti kind of money. I sold a parcel of farmland that I had inherited some ten years ago, and I had this fantasy of investing the proceeds in swampland. Not the classic method of getting rich, I realize, but I wanted my own little corner of black water and gum, a place where the footing is tenuous, a place I could explore at leisure and speculate about its past. I wanted my own real estate of ill repute.

I might as well have tried to buy the Brooklyn Bridge. Carole did find a fifteen-acre parcel of saltwater wetlands priced at twice what I received for the farm. Another tract was more reasonably priced, but it didn't include the cost of the helicopter that I'd need to reach it. Worthless swampland is now known in real estate circles as "waterfront" and is priced accordingly, even if it takes a high tide and a three-day storm to get your feet wet.

After several months, I concluded that it's no longer possible to buy fifty acres of worthless swampland in America, at least not on the mid-Atlantic coast where I live. And this is both good and bad. It's good in that Americans have learned to value swamps and to protect them, although we have devised numerous legal pry bars to move aside legislation that might protect a swamp from, let's say, a planned shopping mall. It's bad in that it seems I'm doomed to do my exploring and ruminating in other people's swamps.

"Other people," for the most part, means "all of us" because the major American swamps (for example, the Everglades, Great Dismal, Okefenokee, and the like) have been set aside by the federal government as parks and national wildlife refuges. Many states have also protected swamps and saltwater wetlands as parks, preserves, and wildlife management areas. Private conservation organizations such as The Nature Conservancy have saved tens of thousands of acres of swamps around the world.

It wasn't always so. The role of the federal government as

protector-of-land is a relatively recent phenomenon. Modern wildlife conservation began in 1908, when President Teddy Roosevelt called a conference of state governors to take an inventory of natural resources. Eight years later the National Park Service was created, and the Migratory Bird Conservation Act was passed in 1929, giving birth to our system of national wildlife refuges.

Until they were protected by federal ownership and legislation, America's wetlands were devastated as the nation grew tremendously in the nineteenth and early twentieth centuries. Over a span of a few generations, we destroyed more than half the swampland that existed when the colonists settled along the Atlantic Coast in the early 1600s. In the Northeast, as population grew, swamps and bogs became especially vulnerable. In 1780 the region had an estimated 80 million acres of various types of wetlands. A 1980 wetlands inventory showed that fewer than 33 million acres remained, a loss of nearly 60 percent.

Why did this happen? For one thing, we simply knew no better. Today we realize that swamps play a vital role in matters such as flood control and water quality. We have come to appreciate swamps as unique places, as homes to birds, animals, fish, and plants common only to this special ecosystem. But two centuries ago, in the exuberance and naivete of a young nation, we had the collective notion that swamps were of no practical value, and in good American fashion we went to work, drained their lands, and converted them to farms and homesites, towns, and cities.

But could there have been more to it than our desire to make something useful out of land we perceived as worthless? Two hundred years ago, when we were draining swamps like mad, we already had more farmland than we needed. And there were all kinds of places to build houses. There is something in us that fears swamps, and it has pervaded our literature, history, and culture. We have the image of Humphrey Bogart trudging through a swamp, covered with leeches in *The African Queen*. Ernest Hemingway equated swamps with madness and death in his short story "Big Two-hearted River." The bad guys always got eaten by crocodiles or swallowed up by quicksand in Tarzan movies. The swamp is the home of the serpent. We walk in the shadow of Swamp Thing.

I find this attitude puzzling. After all, a swamp is by definition a forested wetland. We like forests. They provide trees for building homes and are shady and cool. Deer live in forests; Robin Hood and his Merry Men made their home in one. Winnie the Pooh lived in the Hundred Acre Wood. We also like rivers, lakes, and ponds. We go out on them in boats and refresh ourselves in them when the weather is hot. Yet, if you combine a forest with water, it becomes a swamp, and people fear it.

This is a book about the natural history and human history of swamps. Up to this point, I have used the term "swamp" in a rather general manner. When someone mentions swamp, many of us picture a wet, low, soggy landscape not easily accessed by foot. Depending upon where we live, a swamp could be a bog, a fen, a wetland, a slough, a pothole, a wet meadow, or a marsh.

These terms are not really interchangeable, although each refers to some type of wetland. Bogs are found primarily in the northern United States and are depressions where a moist environment has combined with an accumulation of organic matter, which eventually turns to peat. A fen is similar, and fens and bogs are often found near each other, but a fen's moisture comes from a stream or groundwater, while a bog is fed primarily through precipitation. Potholes are found where thousands of years ago glacial-driven rocks scoured circular depressions in the landscape, which have filled with vegetation. Their moisture comes from groundwater and precipitation. A marsh usually refers to a tidal wetland. The water is salty and the primary plants are grasses such as saltmarsh cordgrass (*Spartina alterniflora*) and saltmeadow hay (*Spartina patens*).

So "swamp" is more generic than such definitive terms—fen, bog, or pothole—all of which could be described as swamps, or forested wetlands. The business of defining wetlands has been a tricky one, especially since 1977, when Congress passed Section 404 of the Clean Water Act, which controls dredging, filling, and otherwise manipulating wetlands. Although the legislation seems very straightforward, the need to define "wetland" created something of a slippery slope. A conservationist's definition of a wetland is not necessarily the same as that of a builder of shopping malls.

The situation was further complicated in 1985, when Congress passed the "swampbuster" provision of the Food Security Act to deny federal subsidies to any farmer who intentionally converts wetlands into farmland. Again, the definition of "wetland" became the pivot-point upon which the legislation teetered, and conservationists, developers, and farmers are still trying to move that point a little closer to their respective sides.

In general, plants and soil type have become the defining factor in determining whether a parcel of land is a wetland. Wetlands have what scientists call hydric soils—that is, soils developed under wet conditions. Only a few plants, known as hydrophytes, have become adapted to living in hydric soils. If hydrophytes are growing on the parcel in question, and if the soil is hydric, then chances are good it's a wetland. (This is discussed further in chapter 3, "Swamps and People.")

While we have passed legislation governing alteration of wetlands, there is no true national law to ensure their protection. Wetlands are still being drained, filled, and paved over, it's just that a permit is needed to do so. According to the Academy of Natural Sciences, 300,000 acres of natural wetlands are lost each year in the United States. Wetlands fall under the jurisdiction of numerous federal agencies, including the United States Department of Agriculture (USDA) Forest Service, the Environmental Protection Agency, and the Army Corps of Engineers, which has the authority to determine which lands are subject to wetlands regulations. Many states also have wetlands regulations.

The government's goal of "no net loss" of wetlands depends greatly upon such concepts as compensatory mitigation, wherein a developer will restore or create new wetlands to compensate for those lost to legally permitted development. But most scientists agree that it is difficult to create a fully functional wetland where none previously existed.

More promising is the effort to rejuvenate wetlands that were altered many years ago. Often, in the case of agricultural land, this can be accomplished simply by removing ditching so the soils can rehydrate and wetland plants can reestablish themselves. Ironically, a lot of the draining of swampland that occurred fifty to one hundred years ago happened with the assistance and blessings of the Army Corps of Engineers, something that the Corps is helping today to undo.

One of the problems with mitigation is its emphasis on numbers. A developer is allowed to destroy swampland here if he agrees to create swampland there. Our no-net-loss philosophy omits a fundamental fact of nature: swamps do not stand alone, but are part of a broader natural landscape that might include rivers and streams, bays, brackish marshes, salt meadows, upland forests, shrub communities, and grasslands. All of these elements function together with some degree of natural overlap. There is no well-defined edge, and efforts to impose one reflect human thinking rather than natural realities.

Wetlands are not equal. If you destroy ten acres of swamp in a fully functioning forested wetland environment, you harm water quality, reduce the ability of the swamp to eliminate excess nutrients and toxins, destroy valuable spawning grounds for fish and shellfish, and diminish the land's ability to control flooding and erosion. You harm not just the swamp, but the broader natural—and human—community that surrounds it. Yes, you could possibly create ten acres of wetlands elsewhere. But it wouldn't be a fair trade.

2

Water

THE SOURCE OF LIFE, THE AVENUE OF COMMERCE

THE FIRST MOMENTS in a canoe are always the best: gliding away from land, buoyant in a small boat, balanced on a fulcrum of water. I'm sometimes reluctant to paddle, to interrupt this feeling of weightlessness, of perfect balance. The initial pull provides surprising speed. The little boat surges ahead almost silently, only a whisper of water as the bow slices through, a paddle-drip on solid black water. The boat turns to the left and I respond by turning my left hand forward and down, lifting with my right, ending the paddle stroke in a "J" motion, bringing the bow back around and keeping the boat on course.

I was on a blackwater stream in southeastern Maryland and it was spring, white clusters of fringe-tree blossoms hanging over the banks, wild azalea in all its glory, pickerelweed and arrow arum emerging, bald cypress sporting lacy young leaves. It took me only a few strokes to leave behind the bridge, the truck, the roadway, civilization. I put down the paddle and lay back, took a deep breath, listened. It is remarkable how quickly a small boat can transport you from the everyday to the extraordinary, from the wants and needs of people to intimate commerce with damselflies and warblers, watersnakes and freshwater mussels.

The water was black and glistening in the sun, mirroring the green canopy of gums, maples, oaks, and bald cypress. The water moved ever so slightly, flowing quietly, flexing like sheet metal. Up in the canopy the young leaves rustled, but down on the blackwater stream life was calm, as if removed from the real world.

I've found that the best boats for exploring swamps are the simplest: small, portable, maneuverable, stable, quiet. I have a kayak, but it seems too high-tech for swamps. I don't need a rudder, spray skirt, storage compartments, foot braces. Besides, after having knee surgery not long ago, I find the kayak difficult to get into and out of. It reminds me of that bulky, foot-to-hip leg brace I had to wear for so long after surgery. It can be a mite claustrophobic.

We also have a well-worn, robin's-egg-blue Old Town canoe that we bought used more than twenty years ago. It's not the perfect swamp boat, but it is a beloved member of the family. It's one of the first fiberglass canoes Old Town manufactured and was made from a cast of their original wood-and-canvas boat, with a molded deck that goes all the way around the gunwale, and no center thwart. Not surprisingly, it is a heavy boat. My wife, Lynn, and I bought it before our son, Tom, was born, and I have pictures of her, ominously pregnant, standing in the bow of the Old Town amid a sea of blooming pickerelweed. Afterwards, there are snapshots of Tom sitting in the center of the boat in a walker, then sporting a Snoopy life jacket as he dines on a peanut butter and jelly sandwich, and still later helping to propel the canoe with a small paddle I carved from a one-by-six left over from a building project. The canoe took us up the Pocomoke, the Nassawango, on many seaside creeks and bays, and on camping trips to Lake Drummond in the Dismal Swamp. Long may it live.

But for exploring on my own, I've settled on another Old Town canoe called the Pack. The Pack weighs a little more than thirty pounds and is ten feet long, beamy and stable, perfect for exploring the sheltered, twisting waterways of swamps. I can slide it into the back of my truck, grab the paddle, life jacket, and lunch, and I'm off to explore. The Pack can be propelled with either a double- or single-bladed paddle. The double is better for wider water when you're intent on getting from point A to point B; the single works better in narrow streams when maneuvering is priority.

The Pack is built for one person. It has a single cane seat, a wooden center thwart, and it is made of Royalex, a lightweight type of plastic. I added a wooden shelf that clamps onto the thwart, with edges that follow the contour of the gunwale, a perfect place to keep at hand your binoculars, field guide, lunch, or anything else you don't want to collect paddle-drip in the bottom of the boat.

The Pack is a small, uncomplicated boat, and it makes me feel at home on the water, especially in these narrow passages of swampland. It allows me to enter the world of the swamp, but it does not intrude.

Water is the key to the swamp. It allows me access. Most of our mid-Atlantic swamps are fed by streams, by runoff from higher land, and these streams provide access to places that would otherwise be beyond reach. Dragon Run, on Virginia's Middle Peninsula, is one of the state's most pristine streams, meandering some thirty-five miles through four counties before it widens near the town of Saluda and becomes the Piankatank River. Eighty percent of the Dragon's watershed is forested, so you paddle through isolated cypress swamps and at least for a few hours enjoy the feeling of wilderness. Streams provide access to the cypress swamps of eastern Maryland, with the Pocomoke River flowing from the Delaware line southward to the Chesapeake Bay. Nassawango Creek leads to a wilderness preserve of 9,000 acres, and the two-mile-long Feeder Ditch runs into Dismal Swamp and the Army Corps of Engineers campground at Lake Drummond.

Over years of exploring swamps it has occurred to me that water is the defining character of the landscape. Obviously, without water there would be no swamp. We destroyed more than half of our native swamplands by depriving them of moisture. We ditched and drained and plowed and filled until they became good solid land, suitable for farming, for subdivisions, shopping malls, airports, and garbage dumps. We know a swamp is defined by the presence of water and trees—a forested wetland. But the more time you spend in swamps, the more you begin to understand what a complex role water plays in the lives of the animals and plants that live there.

The prothonotary warbler is a true swamp bird, and water is central to its life. Indeed, in some quarters the prothonotary is called the swamp canary. It is a stunning bird with a golden-yellow head, breast, and belly, a blue-gray back and wings, and dark, prominent eyes. Prothonotaries are so named because back in the eighteenth century, when the Creoles first saw them in the Louisiana swamps, they thought the bird's bright plumage resembled the golden robes of the protonotarius, an official and scribe of the Catholic Church.

You're not likely to find one of these warblers in upland forests; it truly is a bird that has adapted totally to swamps. It's the only eastern warbler that nests in cavities (most are cup-nesters), and frequently it will make use of the abandoned nests of downy woodpeckers. The prothonotary resides near the water, and sometimes will actually settle in a stump, dead branch, or overhanging cypress knee. The male builds "dummy nests," but the actual nest is built by the female, which lines the cavity with moss, lichen, dried leaves, and strips of bark. By building over the water, the prothonotary escapes many of its predators, and although flooding is a possibility, fledglings have developed the ability, unique among songbirds, to swim.

Prothonotaries are frequently seen in swamps in the southeastern United States from spring through late summer. They begin nesting in April, with the female laying a clutch of four to six eggs. She incubates them for twelve to fourteen days, and the young are ready to fledge eleven days after hatching. In late July to mid-August, the adult pair and the young-of-the-year hatchlings leave the swamp and begin a southward migration that takes them to Mexico and farther south to Columbia and Venezuela, where they spend the winter in mangrove and freshwater swamps. The male and female pair frequently maintain their bond over winter.

While water is essential to the nesting ritual of the prothonotary, it also plays a role in its diet. In the Southeast, the birds can frequently be seen flying just above the water, hovering to dispatch a flying insect or foraging among thickets of pickerelweed or pond lily. They eat ants, mayflies, aquatic beetles, the larvae of aquatic insects, some crustaceans, and snails. And this colorful warbler is the only bird whose song is actually "tweet, tweet, tweet."

On a spring Saturday I was paddling the upper Nassawango near Snow Hill, Maryland. It was sunny and still, trees freshly leaved, shrubs leaning over the creek with shiny new foliage just uncurling. I looked at the forested shoreline, the cypress knees, all reflected in a mirror of black water, and I realized I was witnessing only half of the life of the swamp. I watched prothonotary warblers hunt insects in rotting logs. I surprised a pair of wood ducks, which took off with an awful squeal. I saw a bald eagle soaring overhead, and ospreys, turkey vultures, and a great blue heron. I spotted raccoon tracks in streamside mud, a midden of little mussel shells. I spied a barred owl back in the swamp, perched on a snag about six feet off the ground. I heard pileated woodpeckers calling back in the woods, like the jungle sounds in old Tarzan movies.

But I realized something was missing. I was floating on the surface of a mirror, seeing and hearing only what was above the surface, and what was reflected in it. Beneath me was a totally different world, a world of water and all the diversity of life it supports.

A dragonfly landed on my knee. It was a beautiful insect, glossy black with gossamer wings and a tiny sky-blue patch on the tip of its abdomen. Overhead, its larger cousins hovered like hummingbirds, stalking small flying insects. Dragonflies, and smaller damselflies, seem to me to be the signature insect of the swamp. The adults mate as they fly overhead, and the female deposits her eggs in or near the water. The eggs hatch and the insects live underwater in a nymphal stage, sometimes for years, until they finally climb out on a stick or plant stem, shed their old shell, and become adult insects.

So while adult dragonflies are amazing to watch—they are skilled, and often iridescent, hunters—I know that most of their life is spent underwater, and when I see a brightly colored adult, I'm witnessing the climatic moments of its life. It will find a mate and together they will produce eggs, which will become nymphs and begin the cycle anew. The nymph is a welcome friend in that it feeds on the larvae of other insects, such as mosquitoes and biting flies, that humans find aggravating. And adult dragonflies feed on adult mosquitoes and flies. This is one of the reasons mosquitoes are rarely a problem in many swamps. (These wonders are the subject of chapter 7, "Delicate Damsels and Dragons That Fly.")

What else could be going on under the surface? What is happening in the world of water? In summer, the blackwater streams I explore on the mid-Atlantic are often lined with fleshy green plants, some with colorful flowers like the purple spikes of pickerelweed, or fragile white and yellow water lilies. Seeing these broad leaves and colorful flowers reminds us that there's a lot more going on below.

Plants such as pickerelweed, water lilies, and arrow arum are called emergent herbs in that the roots and rhizomes are below the surface of the water, while other portions arise from the water during growing season. Pickerelweed is one of the most colorful plants in shallow freshwater streams and ponds. It begins blooming in June and holds its color through the summer. Blooms begin at the bottom of the flower spike, with only a few flowers emerging at a time, and move upward as the months wear on. At the end of the flowering season, seeds will have formed along the spike, which bends over and releases them into the water in the fall. Though the seeds germinate during the next growing season, pickerelweed reproduces mainly through its system of buried rhizomes, from which grow numerous shoots.

Like many aquatic plants, pickerelweed leaf stalks have aerated chambers that keep the plant afloat and upright in water. During dry spells, or at low-tide intervals in tidal streams, the plants collapse on the muddy bottom because they have no internal support.

Pickerelweed is pollinated by bees, butterflies, and ruby-throated hummingbirds, and the plants are often a refuge for the aquatic nymphs of dragonflies and damselflies, which climb the stalks when they are ready to shed their exoskeletons and take to the air. Female dragonflies and damselflies will sometimes deposit their eggs on the lower portions of stalks, to be washed into the water after rains or high tides.

And though the seeds are not the plant's primary means of reproduction, they are not wasted. After being deposited in the water in fall, they provide food for such dabblers as mallards, pintails, teal and black ducks. Deer frequently eat the plant, and muskrats sometimes work the stems into their lodges. But despite its evocative name, this weed is not known to have a special association with pickerel or any other fish.

The plant that is probably most associated with freshwater streams, swamps, and ponds is the water lily, another family of emergent herbs with broad floating or emergent leaves and large white and yellow flowers. The most common variety along the mid-Atlantic coast is the white or fragrant water lily (*Nymphaea odorata*), which has large, white, sweet-smelling flowers and plate-shaped leaves with purple undersides that float on the surface of the water. The yellow or bullhead lily (*Nuphar variegata*) has large yellow flowers, and spatterdock (*Nuphar advena*) has yellow flowers and leaves that emerge like broad funnels. And though water lilies go by different names in different geographical areas—including pond lilies, lily pads, water nymphs, water cabbages, alligator bonnets, beaver roots, and cow lilies, among other colloquial terms—the plants we find in mid-Atlantic swamps are one of these three species.

From spring through fall we enjoy seeing their broad leaves and their flowers, but the remarkable thing about water lilies is the structural engineering hidden beneath the surface of the water. These plants grow from thick rhizomes embedded in the bottom mud. Stems shoot up from the rhizomes and break the surface, each with several air-filled passages called *lacunae* that give them buoyancy. Botanists have found that the air therein does not merely allow the plant to float, but helps transport gases between the rhizomes and the leaves.

The wildlife biologist and writer John Eastman calls this phenomenon "internal winds." In *The Book of Swamp and Bog* (published by Stackpole Books in 1995), he observes that the air is under considerable pressure and moves gases (mainly carbon dioxide and methane) from the sediment layer up through the *lacunae* to the older leaves, releasing these gases into the atmosphere through openings called stomata. In younger leaves, the stomata draw air into the leaf *lacunae,* and from there it travels down to the buried rhizomes and roots in the stream sediment. The system, Eastman says, works like a pump; pressure generated by the young leaves travels through the plant and is vented through the older leaves. Eastman notes that one study measured twenty-two liters of air passing through the leaf stalk of a single water lily in one day.

While water lilies are remarkable examples of botanical engineering, they also are insect magnets. The fragrant flowers are pollinated by honeybees and other small bees called halictids, as well as by various flies. Look closely inside new flowers and you may well see drowned insects. Such casualties are attracted to the flexible stamens, which bend inward under the weight of the insects, causing them to fall into the stigmatic fluid, the small bowl of liquid that pools atop the concave stigma.

Longhorned leaf beetles (*Donacia*) spend their entire life cycle on water lily plants. The adult beetles of roughly five species mate on the flowers or leaves and help pollinate the plants. Tiny black insects called thrips live inside the flowers, as do dance and shore flies.

Eastman writes that water lily leaves are one of the most populous microhabitats of any swamp environment: "They host numerous feeders, egg-layers, shelterers, and resters. One need only glance at the often ragged, riddled condition of many of the leaves in midsummer to note how heavily they have been used. Each leaf provides a detailed record of its uses and users, a historical census of its residents and consumers."

While the leaves provide shelter and cover for small fish, they are also landing pads for dragonflies, damselflies, and other flying insects. Beetles eat the leaves, and casemaking caterpillars chew away flaps to fabricate their homes. The submerged undersides are often covered with egg masses of such insects as damselflies and caddisflies.

By the end of the growing season, what were once broad, fleshy leaves are tattered and torn, devoured by everything from caterpillars to beavers, weathered by storms, and beaten by waves and currents. But leaves and flowers are transitory, the parts that show. Beneath the surface, the sun's energy is stored in rhizomes and roots. And next spring, as the water warms again, new green growth will emerge.

When it comes to exploring the life that lies beneath the surface of a swamp, I often use a fishing rod. In early spring, perch migrate up the streams to spawn. I take the canoe up to the headwaters and quietly cast shad darts tied to six-pound test line. Perch are not big fish, but they are entertaining to catch on light tackle, and after a winter spent eating frozen fish, I enjoy taking them home and having them sautéed and seasoned with salt and lemon juice.

Fish are near the top of the food web in the swamp, surpassed only by the ospreys and bald eagles that soar overhead, or by great blue herons that stalk the shallow edges of streams and ponds. And now and again they fall victim to predators casting shad darts.

A healthy swamp is a vibrant, living, self-perpetuating ecosystem. The food chain begins with the sun, and the medium in which it works is water. This energy source, plus all the living and nonliving components that interact with each other to sustain life—plants, animals, bacteria and other agents of decomposition, air, soil, and water—allow the system to function and flourish.

In essence, solar energy is captured by plants through photosynthesis and is gradually passed along to other animals. Let's say a mosquito larva feeds on a bit of microscopic algae. Energy is transferred from the sun to the algae to the larva. Hours later the larva is eaten by the nymph of a blue darner dragonfly. Still later the nymph is ambushed by a bluegill sunfish in search of a nutritious snack. Then school lets out and two ten-year-olds fish with worms and bobbers from a bridge that crosses the stream. I don't have to tell you what's going to happen to that bluegill.

What we see here is a simplified example of a food chain. Sunlight is converted to and stored as chemical energy (simple sugars, or food), and passed from plant to animal to animal. Obviously, most animals have more than one source of food. Bluegills are predators with a wide-ranging diet, and most ten-year-olds are known to be indiscriminately omnivorous.

Simple food chains overlap in numerous layers to form a food web, in which energy is transferred on several levels. Green plants, which produce their own food through energy captured from the sun, are at the first level of the food web. Herbivores consume plants to capture energy, and omnivores eat both plants and animals, and carnivores eat only other animals.

When a plant or animal dies and is not eaten by another animal, it is broken down by decomposers such as bacteria and fungi, which reduce the tissues to simple compounds that can be used by other plants or by filter feeders such as clams and mussels. And so the web continues.

Think of the plants and animals of the swamp as a food pyramid. The producers, the plants, make up the wide base, where most of the energy is derived. Traveling upward we find microscopic animals called zooplankton, and then nymphs and larvae, tadpoles, beetles, water striders, crayfish, freshwater mussels, small fish, big fish, flying insects, small birds, big birds, small animals, big animals.

Scientists call each strata of the pyramid a "trophic" level. The term comes from the Greek word *trophikos,* meaning nourishment. As energy is passed upward from one trophic level to another, it dissipates. Thus fewer organisms can exist at the upper levels than at the lower ones. The energy available at the various trophic levels represents the ecosystem's carrying capacity.

All the components in a healthy ecosystem are connected to each other. Life-giving energy is passed along in an infinitely sustainable cycle, thus making the ecosystem self-perpetuating. In a swamp, energy is derived from the sun, and through the medium of water it is shared by all. Ecosystems suffer when one element of the cycle is changed or eliminated. The most extreme results occur when we deprive a swamp of water, by ditching, damming, filling, or diverting streams.

Without water, hydrophytic plants such as pickerelweed, arrow arum, and water lily dry up and die. There is no longer green algae for larvae to feed upon, no medium to capture the sun's rays and convert them to chemical energy. Dragonfly nymphs have no mosquitoes to stalk, nor do they have leaf stalks to climb when they are ready to take flight as adults. Ten-year-olds have no bluegills to catch. Those of us who enjoy exploring wild places in small boats can leave our canoes at home. We drive to what was the swamp and explore by foot, and don't even get our feet wet. Nothing is as it was. Once destroyed, a swampland ecosystem is likely gone for good. Asphalt is forever.

3

Swamps and People

OF ALL THE natural systems on Earth, none has been quite so vilified as swamps and wetlands. Nationwide, we have destroyed about half the wetlands that were here when the European colonists arrived. In the Northeast, it's even worse. A USDA Forest Service report estimates that in 1780 there were 79,791 acres of wetlands in what is now the twenty-state Northeastern Area, an administrative unit of the Forest Service. The National Wetlands Inventory of 1980 put the figure at 32,818 acres, a loss of nearly 60 percent. And it's a good bet the figure is even lower today.

Let's think of this in terms of some other natural system. What if we had flattened 60 percent of our mountain ranges, and destroyed all of the plant and animal communities that they supported? What if we had drained 60 percent of our natural lakes and killed the resident fish and aquatic plants? Americans would be vilified worldwide as environmental barbarians (as, in some quarters, we are).

The Forest Service's Northeastern Area—which begins in

Maryland, runs northward to Maine, and then westward to Wisconsin, Iowa, and Missouri—doesn't include Virginia, which has the dubious distinction of being the nation's leader in the destruction of wetlands. According to the Forest Service, the Old Dominion lost 9 percent of its forested wetlands in a twenty-one-year period from the mid-1950s to the mid-1970s.

The Chesapeake Bay watershed, which comprises portions of Delaware, Maryland, New York, Pennsylvania, Virginia, and West Virginia, lost 9 percent of its coastal marshes and 6 percent of its inland vegetated wetlands during this period. When most of us think about the bay we picture sandy beaches and lush salt marshes, but its watershed reaches deep into the Northeast Corridor, and most of the associated wetlands are various types of inland ones, aka swamps.

According to Forest Service estimates, 12 percent of the wetlands associated with the bay are estuarine, or coastal. That means the remaining 88 percent are swamps. The Chesapeake is, of course, losing both types of wetlands, and some of this loss is natural. We're in a period of geological flood tide, and the sea level is rising, so what might have been a cordgrass marsh fifty years ago could very likely be underwater today. But most of our coastal wetlands are disappearing due to residential and commercial development and agriculture.

This is not a new phenonemon. America has had an obsession with destroying swamps since the time of our founding fathers. George Washington, hero of the American Revolution and the man who would not be king, wanted to drain the Great Dismal Swamp. I learned in my seventh-grade Virginia history class that draining the Dismal was the first of many heroic acts attempted by our first president. Washington wanted to take what he deemed a worthless tract of wild land and convert it to productive farmland, or perhaps dig canals so he could cut the timber and ship it to market. The father of our country set a precedent that we followed for two centuries. He created a mindset for young America: Let's take something we consider useless, and through diligence and sweat turn it into something that is productive and valuable.

Washington, of course, did not succeed in draining the Great Dismal, although he did leave his mark. He, together with several other entrepreneurs, formed two syndicates that dug a five-mile-long canal from the western edge of the swamp to Lake Drummond at its center, with the blessings of the Virginia General Assembly. That canal is known as Washington's Ditch and is perhaps the first "monument" to our first president. If you stop at the visitor center in the Great Dismal Swamp National Wildlife Refuge, near the city of Suffolk, you'll find Washington's Ditch nearby.

The notion that swamps may have beneficial uses is a fairly recent phenomenon. In Washington's day, and for two centuries thereafter, America was a swamp-busting nation. Draining swamps and converting them to practical use showed our grit and determination, like driving the railroad westward.

Too often, swamps and marshes became convenient dumping grounds. Towns that grew up around wetlands more often than not used them to dispose of their trash. Eventually, after enough trashed was dumped, the wetland wasn't a wetland anymore.

Jamaica Bay in Queens, New York, was essentially an undisturbed wilderness until the late 1800s. A shallow bay that sunlight easily penetrated, producing lush aquatic plants that were spawning grounds for fish, crabs, and other marine animals, it was rimmed by 25,000 acres of salt marsh, where spartinas grew thick in summer, and in winter were broken down by bacteria to form a nutrient-rich soup called detritus, the ecosystem's staff of life.

In the years following the Civil War, Jamaica Bay ranked up there with Great South, Barnegat, and Chesapeake bays in the production of oysters, finfish, and other valuable seafood. In 1878 the secretary of war sought to "improve" the bay as a commercial seaport, a proposal that was later backed by the City of New York. Channels 1,000 to 1,500 feet wide were dredged and the spoil was dumped on the salt marshes, creating uplands where docks and piers could be built.

Later, additional dredging filled in the salt marsh where Floyd Bennett Field and John F. Kennedy Airport were built. As the population of New York grew, so too did the need for places to dispose of solid waste, sewage sludge, and other garbage, and the salt marshes of Jamaica Bay were put to use. By the time we learned the value of wetlands, especially tidal salt marshes, nearly half the original marshes that lined Jamaica Bay had been destroyed. A century of "improvements" had reduced them from 25,000 to 13,000 acres.

The same scenario unfolded in most of our port cities along the mid-Atlantic coast. Philadelphia, Baltimore, Wilmington, and Hampton Roads were all built on marshes. Usually, channels were deepened by dredging, and the dredge spoil was deposited on the marsh, which also became the repository of the community's waste.

And there really were few alternatives. The geology of the unglaciated coast is characterized by sandy barrier islands, shallow bays, and vast salt meadows. There are few examples of high, hard land bordering deep water. Commerce was driven by ships, and ships needed deepwater ports. Destruction of America's wetlands, for the most part, was not an act of willful desecration, but a practical solution to problems brought on by the growth and expansion of a young nation. We simply didn't know better.

By the time we realized that salt marshes and forested wetlands are important, we had destroyed half of them. And then we set out to preserve what was left. In 1938 Mayor Fiorello La Guardia and Parks Commissioner Robert Moses developed a plan to save what remained of the Jamaica Bay wetlands. Moses published a booklet entitled "The Future of Jamaica Bay," which recommended that the city protect "the scenery and waters, preserve wildlife, reduce pollution, and encourage swimming, fishing, and boating."

In 1948 New York City acquired lands along the bay for protection under Moses's long-range plan. Three years later Jamaica Bay was surveyed by the U.S. Fish and Wildlife Service to assess its potential as a wildlife refuge. A report, written by Clarence Cottam, recommended the creation of fresh or brackish impoundments and the planting of native trees, shrubs, and grasses to enhance wildlife habitat.

In 1953 impoundments were built on a large island where the wildlife refuge visitor center is now located. The project came about through a partnership between the New York City Parks Commission and the New York Transit Authority. The transit authority needed sand to create an embankment along the Rockaway subway line, and in exchange for permission to dredge the material from Jamaica Bay, it agreed to build dikes to create two freshwater impoundments, known today as East and West ponds.

After the impoundments were formed, the horticulturist Herbert S. Johnson was appointed refuge manager, and he set out to implement the programs that Cottam had recommended. Native trees and shrubs were planted, many of which Johnson propagated in his own backyard. Habitat enhancement continues to this day, with plants being propagated at the park's greenhouse at Floyd Bennett Field for use elsewhere.

Under Johnson's leadership, the Jamaica Bay refuge became one of the most important migratory bird sanctuaries in the Northeast. In 1972 Congress established the Gateway National Recreation Area, which included the Jamaica Bay refuge, and directed the secretary of the interior to "protect the islands and waters within the Jamaica Bay unit with the primary aim of conserving the natural resources, fish, and wildlife located therein and permit no development or use of this area which is incompatible with this purpose."

Breezy Point is a narrow spit of sand that separates Jamaica Bay from the Atlantic Ocean. In spring you can see endangered piping plovers and least and common terns nesting in the shell litter on the beach. They are joined by red knots, sanderlings, and a variety of sandpipers that are just passing through, pausing to refuel before resuming their migration to nesting areas farther north.

Watch the coming and going of the beach birds, listen as the noisy conversation of terns fills the air, and you begin to realize what a remarkably resilient and prolific natural system this is. Forget the terns for a moment and look to the north, and you'll see the Manhattan skyline, punctuated by the spire of the Empire State Building. In the foreground, a little to your left, is Brooklyn, and the huge Ferris wheel that rises across the inlet is a Coney Island landmark. Glance to the east and you'll see planes taking off and landing at John F. Kennedy International Airport, one of the busiest in the world.

Surely this is one of the most incredible juxtapositions in nature. You stand amid a setting that looks thoroughly natural and wild, where the human hand has seemingly touched only lightly, yet just beyond is one of the busiest centers of commerce in the world.

If this is a testimony to the resilience of nature, it is also is a confirmation of our need to embrace an untrodden portion of

Earth where the birds and animals and fish and insects carry on as they always have, and always will. Today we realize that open spaces such as these are as necessary as the skyscrapers, offices, freeways, and airports. Cities fuel the human need to work and communicate; wild places such as Jamaica Bay and Breezy Point are equally essential to the human spirit.

What's so great about swamps? A swamp is a landscape that can't decide what it wants to be. It's not water and it's not land, but a little of both. Most swamps are too wet for hiking and too woodsy for boating. So what good are they?

Let's begin with some practical functions. If the Earth was a human body, swamps would be the kidneys, the natural filters that remove impurities from the water and then recycle it into the aquifer or waterway in a cleaner state.

Wetlands do this by entrapping sediments, nutrients, and toxins, breaking them down to simpler and less harmful components. When it rains, water flows into wetlands from parking lots, farm fields, and residential lawns. Pollutants attach themselves to sediment particles suspended in the runoff, and are deposited with the sediment when the liquid enters a wetland.

Once this happens, the pollutants can be broken down by bacteria and used for food by other organisms, or taken up by plants and added to wetland sediments when they die. This biological or physical entrapment improves water quality downstream, and the filtering ability of wetlands similarly protects groundwater by removing contaminants before they seep into the aquifer.

The practice of channelization eliminates, or greatly reduces, this filtering ability. Instead of being entrapped, water rushes downstream with its load of sediments, and the attached toxins go along for the ride. A major reason the Chesapeake Bay suffers from excess nutrients and other pollutants is that many of the small tributaries were channelized, or ditched, many years ago. Couple this with the fact that thousands of acres of wetlands were converted to agricultural, residential, and commercial use, and the bay's natural filtration system is no longer working as it should. The Chesapeake Bay is experiencing something akin to kidney failure.

Swamps also help reduce streamside erosion and flooding. Wetlands act as natural sponges, soaking up water and temporarily storing it during heavy rains, thus reducing the amount of water moving downstream during a particular time period.

Natural wetlands also play an important role in the economy, especially in coastal areas. They are crucial to the survival of many types of plants, animals, and fish. More than 70 percent of commercially important finfish and shellfish depend upon tidal wetlands during some part of their lives. Blue crabs mate, lay eggs, and produce millions of larvae in tidal marshes and shallow bays. Fish spawn in bays and tidal creeks. And clams, mussels, and oysters continually rely on the shallow, rich waters of estuaries.

Swamps also play an ever-increasing role in nature-oriented tourism. Bird-watching has become the fastest growing avenue of outdoor recreation, and swamps are vital to many species of birds. Some stop during migration in spring and fall, while others stay year round. Birds flock to swamps, and people flock to see them. Ecotourism is a low-impact, sustainable industry that benefits a wide range of businesses, including canoe rental and sales shops, bed-and-breakfasts, tour operators, restaurants, and gift shops. People spend an estimated $10 billion a year to watch and photograph birds that are dependent upon wetlands. Ecotourism has increased visitation at wildlife refuges, meaning more employment opportunities and more money flowing to county coffers from payments in lieu of property taxes.

Beyond ensuring water quality and providing flood protection, economic benefits, and recreational opportunities, swamps play another important role, one that is difficult to measure. An increasing number of us are discovering that swamps are an aesthetically pleasing landscape. No other natural system is quite so diverse, with such a wide variety of plants, birds, animals, and insects. It's true that most swamps are too wet for hiking and too woodsy for boating. But if you can find a stream where you can paddle your canoe, or happen upon a boardwalk trail in a park or refuge, you're in for a wonderful experience. The fact that swamps are not easily accessed makes them even more mysterious and rewarding to visit.

What exactly is a swamp? How do we define a wetland? These questions became important in 1977, when Congress passed Section 404 of the Clean Water Act to establish controls over dredging and filling and authorize the creation of the National Wetlands Inventory, and again in 1985, when it passed the "swampbuster" provision of the Food Security Act to deny federal subsidies to any farmer who intentionally converts wetlands into farmland. During the National Wetlands Policy Forum in 1988, a policy of "no net loss" of wetlands was instituted.

Problem was, no one seemed to agree on exactly what a wetland is, so enforcement of Clean Water laws became complicated. Many federal agencies are involved in wetlands enforcement, and so there have been numerous attempts to define them. In 1956 the U.S. Fish and Wildlife Service developed one of the earliest definitions as part of a wetlands classification system for describing waterfowl habitat. This classification, called the Circular 39 definition, is still used to describe wetland types for wildlife habitat purposes.

Circular 39 defines wetlands as "lowlands covered with shallow and sometimes temporary or intermittent waters . . . Shallow lakes and ponds, usually with emergent vegetation as a conspicuous feature, are included in the definition, but the permanent waters of streams, reservoirs, and portions of lakes too deep for emergent vegetation are not included. Neither are water areas that are so temporary as to have little or no effect on the development of moist-soil vegetation" (this according to Samuel P. Shaw and C. Gordon Fredine in *Wetlands of the United States: Their Extent and Their Value for Waterfowl and Other Wildlife,* published in 1956).

Following passage of the Clean Water Act, a more comprehensive and scientific definition was needed as a tool for the National Wetlands Inventory. The U.S. Fish and Wildlife Service came up with this definition in the *Classification of Wetlands and Deepwater Habitats of the United States* (by Lewis M. Cowardin and others, published in 1979): "Wetlands are lands transitional between terrestrial and aquatic systems where the water table is usually at or near the surface or the land is covered by shallow water . . . wetlands must have one or more of the following three attributes: (1) at least periodically, the land supports predominantly hydrophytes (plants that grow in water); (2) the substrate is predominantly undrained hydric (moist) soil; and (3) the substrate is nonsoil and is saturated with water or covered by shallow water at some time during the growing season of each year."

Gradually, vegetation became the meter by which wetlands were measured. The Environmental Protection Agency and the Army Corps of Engineers agree on the following description, from the *Corps of Engineers Wetlands Delineation Manual* (1987), that is used to legally define wetlands under the Clean Water Act: "The term wetlands means those areas that are inundated or saturated by surface or ground water at a frequency and duration sufficient to support, and that under normal circumstances do support, a prevalence of vegetation typically adapted for life in saturated soil conditions. Wetlands generally include swamps, marshes, bogs, and similar areas."

While it is necessary to have a legal definition of wetlands for regulatory purposes, the USDA Forest Service makes a good point in its publication *Forested Wetlands—Functions, Benefits and the Use of Best Management Practices:* "Wetlands are more than physical places where water is present and certain plants grow. Wetlands perform a variety of unique physical, chemical, and biological functions which are essential to the health of the environment and valuable to society, but which are also difficult to define or identify for regulatory purposes."

Wetlands are hard to pin down because they are both terrestrial and aquatic, but not wholly either. Wetlands tend to have hydric soils—that is, soils that developed under wet conditions. Water is usually present in the root zone, and frequently found on the surface.

Only certain types of plants, called hydrophytes, can live in hydric soils, adapting to a world where water saturates the root zone either permanently or for extended periods of time. On coastal salt marshes, the spartinas thrive in an environment that is both wet and salty. Such conditions would quickly kill other plants (a point that I take up in chapter 5, "The Green Sea"). In brackish and freshwater environments, we find plants such as pickerelweed and water lilies, whose roots and rhizomes live in the underwater sediment layer.

Wetlands differ from uplands not only in their types of plants, but also in the appearance of their soils. All soils are porous, meaning that they have spaces between the grains, but if you look at a range of soil types with a magnifier, you will see a number of distinguishing characteristics. Sandy soil has large grains with large spaces between them, through which air and water can easily permeate and move. Clay soil, however, has small grains tightly packed together and does not drain well.

We usually think of swamps as having dark, rich soil, often black or dark brown in color. The reason for this appearance is the presence of water. In upland soil the spaces between the grains are usually filled by air, and this allows oxygen to react with the elements within. A product of this reaction is color, resulting in soils that are red, orange, yellow, or tan. In swamps the spaces between the grains are filled by water for long periods of time, and the chemical reaction between oxygen and the soil's elements is reversed, producing soil that is black, dark brown, or bluish-gray.

These fundamental differences between upland and wetland soils and plants have made it difficult for concepts such as wetland mitigation to succeed. As I noted in chapter 1, mitigation began as a process whereby a developer would create wetlands to replace those lost in the course of a building project. The word mitigate comes from the Latin *mitigare,* meaning to soften, and to make less harsh, and the intent of environmental mitigation is to soften the effects of development on wetlands and other natural communities. Depending upon whom you ask, mitigation is either a resourceful means of achieving the national goal of "no net loss" of wetlands, or a handy way for developers to skirt environmental laws.

The original concept of mitigation—building new wetlands to replace those damaged or destroyed by development—did not always succeed. These complex natural systems have evolved over thousands of years, and although you can design a landscape to replicate a wetland, there is no guarantee it will remain one ten, fifty, or one hundred years from now.

The process of mitigation has grown to include the restoration of former wetlands, and even the possibility of earning credits by restoring or creating wetlands and "banking" those credits for future use. A mitigation bank is established when a government agency, a corporation, or a nonprofit organization restores, creates, or enhances a wetland area, and then sets it aside to compensate for future wetlands development, under a formal agreement with a regulatory agency such as the Army Corps of Engineers. The value of a bank (its "credits") is determined by the area and quality of wetlands restored or created.

The Environmental Protection Agency (EPA) is promoting mitigation banks, claiming that they reduce uncertainty and delays for qualified projects, especially when such projects are associated with a comprehensive planning effort. According to the EPA, successful mitigation can be ensured since the wetlands can be functional before a project begins. In addition, banking eliminates the temporary loss of wetlands that occurs when mitigation is initiated during or after the impact of development. Consolidation of numerous small, isolated, or fragmented mitigation projects into a single large parcel may have greater ecological benefits. A mitigation bank can bring together scientific and planning expertise and financial resources, thereby increasing the likelihood of success in a way not practical for individual efforts.

Mitigation banking has even attracted entrepreneurs to create or restore wetlands, build up mitigation credits, and then sell them to developers, who consider their purchase part of the cost of doing business, like paving a parking lot. Many developers find that it's much easier to write a check than to get in the business of wetland creation.

The mitigation process in Virginia has been fine tuned, and many of its earlier limitations are currently addressed through a program called the Virginia Aquatic Resources Trust Fund, a mitigation partnership administered by The Nature Conservancy (a nonprofit land conservation organization) and the Norfolk District Corps of Engineers, which is responsible for enforcing federal wetlands legislation. The Virginia Department of Environmental Quality enforces Virginia Water Protection regulations and consults with the Army Corps on project approval for the trust fund.

Rather than building or restoring wetlands as mitigation for altering existing ones, the program allows a developer to pay into the trust fund, which then underwrites the cost of establishing wetlands elsewhere. Generally, the trust consolidates money from many projects with small impacts and pools these funds to undertake larger projects that have a greater chance of ecological success.

In many cases, the trust fund pays for the cost of purchasing, restoring, and monitoring land that fifty or more years ago contained wetlands, but was ditched, drained, and converted to farmland long ago.

Since 1995, 364 projects have used the fund as mitigation for permitted impacts on 147 acres. More than $11 million has been paid into the fund, reflecting an average mitigation value of more than $75,000 per acre. As a result, hundreds of acres of wetlands have been restored, many in sensitive areas such as the Northwest River corridor in southeast Virginia, where The Nature Conservancy has been working to protect endangered plant and animal species.

Marshyhope Creek flows north to south down the Eastern Shore of Maryland, just west of the Delaware state line. It begins northwest of Harrington, Delaware, where it is known as Marshyhope Ditch. It enters Maryland and meanders south through Idylwild Wildlife Management Area and the town of Federalsburg, and then joins the Nanticoke River at Walnut Landing, flowing west to Tangier Sound and the Chesapeake Bay.

Marshyhope seems a lazy country stream, tidal but fresh, a good place to catch largemouth bass and catfish. But looks are deceiving. Like many placid waters, it can become easily riled when too much rain falls in too brief a time. On the heels of a coastal storm in September 1935, the Marshyhope rose seventeen feet, inundating Federalsburg, which lies on the western bank. Homes were flooded, businesses lost, streets turned into canals. But the floodwaters eventually receded, the Marshyhope returned to its banks, and the people of Federalsburg rolled up their sleeves and went to work to restore the town. When the town hall on Main Street was rebuilt, a beige brick was set about four feet above the sidewalk to mark the height of the floodwaters.

Marshyhope Creek and the town of Federalsburg sparred numerous times over the years, and in 1968 the town decided to tame its moody neighbor. Federalsburg and the U.S. Soil Conservation Service devised a plan to straighten and deepen the creek, ensuring the rapid and controlled movement of potential floodwaters through town. So the heavy equipment came in, the creek was dredged and widened, and the swamp forest and marshes that lined it were built up with spoil pulled from the creek bottom. The original river channel was drained and filled. When the project was finished, the Marshyhope ran straight and true from one end of town to the other, a distance of a mile and a half. It was widened by about one-third, to 160 feet, with an average depth of eight to ten feet. The dredge spoil was mainly sand and gravel, with little organic matter, and was used to line the banks of the creek.

While channelizing the Marshyhope alleviated Federalsburg's flood problems, the environmental impacts were considerable.

Where there once were swamp forests and wetlands, there now were sandy, grassy fields, little used by people or wildlife. The Marshyhope's banks were largely barren for nearly thirty years, until the town decided in the mid-1990s to capitalize on its riverfront location and beautify the creek. The Maryland Department of Natural Resources (or DNR), together with the town, came up with a plan to dress up the stream banks and restore the filled swamp forest and wetlands, thereby creating a greenway that would salvage wildlife habitat and create an inviting setting for visitors.

It was an ambitious plan, but could it be done? Could a massively altered streambed be returned to its original condition, or at least be restored to a functioning natural system?

In the summer of 1997 the heavy equipment moved back in to Federalsburg, this time operated by a phalanx of soldiers from the Sixth-Eighth Engineering Battalion of the Maryland National Guard. The guard set up a command post, medical tent, commissary, and equipment repair facilities, and then went to work, removing 50,000 cubic yards of dirt that had filled swamps and streambeds nearly thirty years earlier.

Kevin Smith, now the chief of restoration services with the Maryland DNR, was there when the guard went to work and describes the process as "organized chaos." The guard spent two summers in Federalsburg, working in intense heat to remove fill down to the original level of the old marsh. Some of the logs and stumps that had been buried for thirty years were excavated and placed back in the marsh when the grading was finished.

"The guard removed roughly four feet of earthen fill, mostly sand and gravel, and placed it in an abandoned sand mine adjacent to the site," says Smith. "They did a great job of restoring the site to the original elevations, but when they finished, the area resembled a moonscape. There was nary a blade of grass nor a sprig of sedge to be found."

The subsequent greening of Federalsburg involved numerous community partners, including volunteers from the Chesapeake Bay Foundation, students from Federalsburg Elementary School, and members of the Maryland Conservation Corps, Boy Scouts, Girl Scouts, and the Rotary Club. "We explored pristine sites on the Marshyhope by canoe and we noted plant species and their locations in relation to tidal flows, and elevations were recorded," says Smith. "Also recorded were the types of plants that live together. For example, pumpkin ash seems to occur with seaside alder, the alder preferring a slightly lower landscape than the ash. So the communal associations were used to develop the planting plan for the site."

Once the landscape plan was established, volunteers took to the swamp to collect native seeds. In September, when the wild rice seeds were mature, Smith and his group of volunteers canoed into rice marshes, bent the plants over the gunwales, and beat the seed heads with sticks, releasing wild rice into the bottoms of the boats. All of the varieties of seeds gathered were stored for future use, with the alder and ash propogated at the DNR's John S. Ayton Tree Nursery in Preston and grown at the elementary school in Federalsburg. Then came the task of planting the trees, adjusting tidal flow to fill the newly restored marshes, and waiting for nature to reclaim its own.

Kevin Smith and I walked the Federalsburg Greenway trail on a September morning nine years after the National Guard had rearranged the landscape, and after a corps of volunteers had gathered and propagated seeds and started the process of restoring the Marshyhope. The creek still runs straight and true, but portions of the old streambed are now oxbow ponds, connected to the creek by a small stream, fed by the tidal flow. We stood on a footbridge and watched largemouth bass hold steady in the current, waiting for a meal to come along. Pickerelweed grew along the edge, and we spotted cardinal flower, pumpkin ash, alder, and spatterdock, all native plants that had reestablished themselves. A turtle sunned itself on an exposed log, and butterflies fluttered among wildflowers.

"This is what we've been looking for. This is what it's all about," said Smith, who darted down a streamside slope, grabbed a handful of tall grass, and shook it into his other hand. He held out a few dozen small brown grains of wild rice. "This is what we've been waiting for, to see plants like this come back. It's a signature species for freshwater wetlands like this, very important food for wildlife. When we dug this out, the soil was very sterile, sand and gravel. But wild rice needs an organic substrate, and to see it here now means that the wetlands are functioning as they should. Plants die back in the winter and they add organic material to the soil. This is the first year I've seen wild rice here since the project. That's exciting."

I asked Smith if he considered the restoration project a success. "It would be an overstatement to say that the impacts of channelization have been rectified by the restoration project," he said. "But I think we've alleviated some of the environmental costs, and that is what restoration is all about. We can't return to the days when Captain John Smith sailed up the Chesapeake, but we can work to reclaim some of those things that made the Chesapeake Bay region so wonderful in the first place. Restoration projccts like this are a step in the right direction."

4

Life Begins Here and Goes Everywhere

SO WHERE, then, does a river begin? How does the flow start? Is there some secret spring where water emerges from the earth, beginning with a trickle that deepens and widens, scouring its banks, flooding swamps, and finally asserting itself and gaining strength until people look at it and say, "That's a river"?

The Pocomoke is not a major river. It doesn't divide a continent like the Mississippi. It has no whitewater rapids. It has not carved its way through canyons, and, to my knowledge, it has not been written about in novels. Nonetheless, it is a serene and beautiful river, flowing south and west from a huge cypress swamp in southern Delaware. On its upper reaches the trees lean over it from both banks, forming a canopy that turns a summer canoe trip into a drift through a shimmering green tunnel.

At the town of Snow Hill, in Maryland, the river widens, and by the time it reaches Pocomoke City it is a broad, busy avenue of commerce, where Chesapeake Bay deadrise workboats patrol the flats in search of blue crabs. Thereafter it enters the Chesapeake Bay, but even then continues to assert itself, forming a deep underwater trough called Pocomoke Sound. When fishermen leave ports such as Onancock, Virginia, to go in search of gray trout and croakers, they'll use depth finders to locate productive fishing areas. At the mouth of Onancock Creek you steer a course of about 340 degrees, and for more than a mile the depth will not vary much from twenty feet. But suddenly, just as the channel marker at the mouth of the creek fades from view, the depth drops to thirty, forty, fifty feet. This ledge, known to anglers as Ditchbank, marks the remains of the Pocomoke River. In another age, when glaciers ground their way through New England and into New York, there was a river here, not a bay, and where we fish for trout, men once hunted deer and bear.

So the Pocomoke begins as a trickle, crosses the state of Maryland, and ends up as an underwater ravine flowing beneath the Chesapeake Bay. This odyssey covers fewer than one hundred miles, through a landscape that ranges from cypress swamp to saltmarsh cordgrass. But where, I wondered, does the Pocomoke actually begin? Is there some spot, some crease in a cypress swamp, where we can say, "This is the source; This is where it all starts"?

I have canoed much of the upper portion of the river, from Whiton Crossing to Porter's Crossing to Snow Hill. But I wanted to go farther upriver, to find the source. To do so, I left the canoe at home and went by foot and by car. The best way to see the upper portions of the Pocomoke is to find a country road that crosses it, then park the car and explore by foot.

Upriver from Snow Hill, the first crossing is Porter's. You take MD 12 north out of town, turn right on Whiton Road, travel about five miles, and then turn right on Porter's Crossing Road. It's a narrow, winding farm road that crosses fields, enters a swamp, and crosses the river, and is a popular put-in for canoeists. Back in Snow Hill, Barry Laws's Pocomoke River Canoe Company has trailers filled with red Old Towns, and during the warmer months they make several trips a day, dropping off weekend explorers who will put in here and paddle the five-or-so miles back to Snow Hill.

There's a good reason Porter's Crossing is popular. The five miles between here and Snow Hill are probably the most beautiful on the river, especially if you're paddling a small boat. The Pocomoke is no more than sixty feet wide here. Above, the ceiling is various shades of green; below, the scene is reflected in the dark water.

The river is narrow, winding, and has numerous deadfalls that make it unnavigable for motor boats, so the sounds you hear are natural ones: the cackle of pileated woodpeckers, the squeals of wood ducks taking flight, the "who-cooks-for-you?" chant of barred owls. In spring and early summer the swamp is filled with migrating warblers, staking out territory and communicating with mates, their sounds and songs echoing in the trees. Most obvious is the prothonotary warbler, a brilliant gold-and-gray bird that flashes across the river like sunshine, searching for a cavity in which to nest. After a couple of miles, the river widens, motorized traffic increases, and although the waters are still beautiful and appealing, the fleeting illusion of wilderness swamp wears away.

Five miles northeast of Porter's Crossing is Whiton Crossing. Whiton Road (MD 354) roughly parallels the Pocomoke through much of Maryland, and a right turn onto one of the narrow back roads will usually take you through the swamp and across the river. Whiton Crossing Road intersects Whiton Road in the little community that bears the same name. In a mile or so the swamp appears, and a small wooden bridge crosses the water.

The Pocomoke takes on a different look here. The swamp is

filled with cypress and black gum trees, and the river is covered by a canopy of green. But unlike Porter's Crossing, here the river does not meander. Stand on the bridge, look at the course, and you see something akin to a railroad right-of-way, arrow-straight in both directions, in summer infinite green carved through the heart of the swamp.

It doesn't take a visitor long to realize this is not a natural phenomenon. Nature rarely puts much truck in straight lines, especially in swamps, where water follows the path of least resistance, where trees wrestle for sunlight. Here the Pocomoke was straightened in the 1920s and '30s to improve flow and help drain the upriver swamps.

I parked the car along the side of the road and began exploring the swamp on foot. Whiton is a farming community, with well-kept homes, barns, grain silos, and tractor sheds. It's a neat, clean landscape, where people obviously take pride in where they live. Unlike some rural areas, there was virtually no roadside trash, no ancient Ford Pinto with wild cherry saplings growing through the floorboard.

In the swamp, though, things were different. During a short walk I spotted one broken storm door, three discarded tires and wheels, plastic buckets, and more cans and bottles than I could count. Downriver from the bridge a tree had fallen, and its branches had trapped a surprising clump of flotsam—lightbulbs, a soccer ball, various cans and bottles, a drywall bucket, and a large animal skin, presumably the hide of a deer. Later, while exploring farther upriver, I came across several deer carcasses that had been disposed of along the river; one skin apparently had been caught by the tide.

And it occurred to me that our collective attitude toward swamps has not changed much in the past century. In a landscape that otherwise is kept squeaky clean, trash is inevitably dumped in the wetlands. They are used as a slaughterhouse, a place to dispose of whatever society no longer needs. They are a dumping ground.

The curious thing is that I saw no trash at all at Porter's Crossing. Is it because more people use it as a canoe put-in, and illegal dumpers are more likely to be seen? Or could it have something to do with the altered state of the river? Here, the Pocomoke resembles nothing so much as a large ditch. Spoil-banks line both sides, remnants from when its course was deepened and straightened many years ago. It has an industrial look; it flows with a purpose from here to the horizon. Unlike at Porter's, where the water quickly disappears into a fold of green, there is no mystery. Could this be like an urban building, where one broken window invites others, until in time the building, or the river, is desecrated?

The Pocomoke was channelized from about a mile south of the Whiton Crossing bridge all the way into Delware, where it helps drain the Great Cypress Swamp that straddles the Maryland-Delaware line. I walked along the edge of the swamp at Whiton and found what appeared to be the original streambed, about a quarter-mile west of the channelized course. The water was still and shallow, but the pathway it took through the swamp was easy to see, and it made me wonder what the Pocomoke might have looked like in its original state.

I drove back out to MD 354 and headed north. A right turn on MD 374 took me to Burbage Crossing, where a concrete-and-steel bridge crossed the water. I stood on the shoulder of the road along one bank and paced off the width of the river. Thirty paces. About sixty feet, same as at Whiton. Here, too, the Pocomoke ran straight and true, slowly carrying its load southward in a languid current.

Back on the road to Powelville, another right turn took me to Purnell Crossing. The river is narrower here, with a steep bank, and from the color of the water it obviously carries a heavy load of sediment. Near Snow Hill, the water is dark from tannic acid, but nonetheless surprisingly clear. This water was like coffee with a heavy shot of half-and-half, creamy as café au lait. I parked in a power-line right-of-way and walked down to the river. Here too, someone had brought their deer to be butchered, and rotting skins and carcasses were left along the riverbank.

Farther north, the Pocomoke flows under U.S. 50, which connects the Ocean City resort area with the rest of the world. I stopped in Willards, checked the topographic map, and headed for Delaware, realizing I had lost the river. I headed up to Gumboro on MD 26/30, and on the map the Pocomoke resembled varicose veins, a cluster of blue lines. The road ran through huge farm fields, cut with deep, water-filled ditches. I knew these would eventually join and form the northern portion of the Pocomoke. They were industrial-strength ditches, designed to quickly remove rainfall, and some, like Cowhouse Branch, were actually identified by road signs.

This is farming country, and poultry country. The two are closely linked. The fields produce corn and soybeans, which are blended into feed. You can't drive far without seeing chicken houses, although that term doesn't do them justice. The newest ones are computer operated, and the birds are fed, watered, and tended by a ten-gigahertz mother hen.

The waste from the poultry farms is recycled as fertilizer. I was there in late winter, and mounds of it were piled up, awaiting the spreader trucks. In fields where the manure had already been dispersed, vultures had gathered to feed. Two mature bald eagles were among them. The aroma was distinctive. I once knew a poultry grower pretty well. He was a funny, blunt man who had once been interviewed in a newspaper about his operation.

When asked about the odor, his reply was typically succinct. "Smells like shit to you. Smells like money to me," he said.

According to the topo, the Pocomoke begins with two forked streams just outside the community of Pusey Crossroads, about five miles north of the Maryland-Delaware line. So the Pocomoke's source, according to the map, is two ditches that drain a pine forest in southern Delaware. Those ditches, which the topo identified as the Pocomoke River, are soon joined by other modest waterways—Cowhouse Branch, Cypress Farms Ditch, Bald Cypress Branch, Gum Branch, Gray's Prong, Green Run, Whaleyville Branch, Burnt Mill Branch, Franklin Branch, Timmonstown Branch, Savanna Branch, Ninepin Branch, Duncan Ditch, and Tilghman Race—and by the time the modest river slices its way through the countryside to Whiton Crossing, it has drained more than 1,000 square miles of forest and farmland across two states.

To drain land, of course, is the reason the upper portions of the Pocomoke were channelized years ago. Many of the farms within the watershed were once part of the vast swamp that buffered the river, and hastening the drainage of rainwater meant that its rich, peaty soils could be used for agricultural purposes. The swamps that line the river are noticeably narrower along the channelized portions than where the Pocomoke follows its natural course.

The Pocomoke watershed is but one of several in southern Delaware and the Eastern Shore of Maryland. Many years ago, this area would have been a vast swamp, perhaps even an extension of the Great Dismal that covered much of southeast Virginia and northeast North Carolina. The two landscapes have much in common, especially with regard to botany. For example, the bald cypress is common in southern hardwood swamps such as the Dismal, and is at its northern limits in Delaware.

Remnants of this great swamp are easily seen in places such as

Trap Pond State Park in Delaware and along Nassawango Creek, which joins the Pocomoke River near Snow Hill. The area is, of course, separated from the Great Dismal by the Chesapeake Bay, but if we use our imaginations and go back to the last Ice Age, when the bay more closely resembled a river, then the connection with the Dismal seems very possible.

Even the name "Pocomoke" could have a link with the swamps of North Carolina. Conventional wisdom holds that the word is a Native American term meaning "black water," which in this instance certainly makes sense. But it could also be a variant of the term "pocosin," meaning "swamp on a hill" or "broken ground." The term was used by American Indians to describe the Carolina bays, or elevated swamps, that are common in eastern North Carolina.

John McCoy was standing in the corner of a farm field just south of the Delaware line. McCoy runs the Management Studies Program for the Maryland Department of Natural Resources (DNR), and it is his job to keep tabs on what the water carries from these fields to the Pocomoke River, and eventually to the Chesapeake Bay. We were at an intersection of Gumboro Road and a branch of Green Run, where a concrete bridge crosses the ditch, ten feet wide and two feet deep with a surprisingly swift current. Little sandbars had formed where the current slowed, and in places the stream had undercut the bank.

McCoy is a friendly, stocky man with a bushy mustache. His background is in science, but he often finds himself torn between the interests of farmers and the pressing need to clean up the Chesapeake Bay. One of the major problems he encounters is the effect of excess nutrients such as nitrogen and phosphorus, which are beneficial and necessary in limited doses but can otherwise trigger algae blooms and create oxygen-depleted "dead zones" that rob the bay of valuable fish and shellfish. Bay advocates say that the widespread use of chicken manure on farm fields far inland is one of the major sources of such environmental excesses in the Chesapeake, some seventy miles away.

Chicken producers, many of whom also are grain farmers, clean their poultry houses and truck the manure to their fields, where in early spring the nitrogen will nourish a summer crop of corn. In most cases, the practice would be a wonderful example of recycling; a waste product is used to fertilize a crop that will eventually provide feed for the poultry that produced the waste.

Yet many argue that too much nitrogen is leaching from the fields, or being swept away by rains, and is entering the river and eventually reaching the bay. The Maryland DNR has established Best Management Practices designed to curb the amount of nutrients that enters the watershed, but the practices are voluntary, and it's difficult to convince farmers to change the way they work if it's going to cost them money.

"Farming is a business, and a farmer is not likely to voluntarily adopt a practice if it's going to put him at a financial disadvantage," said McCoy. This means that in many cases, the state helps out. We drove through the countryside and he pointed out a number of places where DNR had worked with farmers to reduce nitrogen runoff. It was late winter and most of the fields wore the stubble of last year's corn or soybean crop. A few were green, with a seasonal planting of annual rye grass. The grass, which prevents runoff and traps and releases nitrogen in the spring when it is tilled, was seeded by the state with the farmer's permission.

On a number of farms, manure was stored under large sheds with sloped roofs and open sides. These protect the natural fertilizer from rainwater and prevent it from leaching into the soil or being washed into ditches. Still, many farms had large mounds of poultry waste piled just a few feet from large ditches.

"It usually comes down to economics," McCoy says. "If a

farmer stores his manure in a shed, it means he'll have to handle it three times. Once to clean out the houses and move it to the shed, then to move it from the shed to the field, and then to load it into the spreader and apply it to the fields. If he doesn't use the shed, all he has to do is pile it in the field, and then spread it when he's ready. So, when it comes to fuel and labor costs, it's more expensive to store manure in a shed."

The state encourages farmers with incentives, such as covering the costs of the sheds. "A carrot on a stick," says McCoy. "We've been monitoring two watersheds in Green Run since 1994, and in 1998 we began an incentive program, encouraging farmers to move their manure out, replace it with inorganic nitrogen, and plant cover crops. It was strictly voluntary, but we had a high rate of participation."

McCoy says a pound of inorganic fertilizer produces a pound of available nitrogen, but when poultry litter is used, it takes twice as much to provide a pound of available nitrogen. "If a farmer needs to add 150 pounds of nitrogen, he'll have to put on about 300 pounds of poultry litter. So if we take the poultry litter out, we cut the nitrogen budget in half, and then we use the cover crop to hold it in the root zone.

"When a farmer goes about his annual nutrient management planning, he'll set a yield goal, let's say 150 bushels of corn per acre, and he knows he'll need 150 pounds of nitrogen to do that. So if he puts that on and we have a dry year, and his yield is only 90 bushels, then that means 60 pounds of nitrogen are still out there. And that's where the cover crop comes in. It locks the nitrogen in the root zone, and it also provides some protection against runoff."

McCoy says problems arise when farmers and poultry producers dump manure on fields just to get rid of it. "There's an extreme difference between waste disposal and nutrient management," he observes. "You can't just get rid of material without paying attention to its nitrogen value, and, truthfully, most people don't do that. Nitrogen is valuable. And the people who farm here live in the community, and I think that for the most part they're concerned about where they live."

Along one ditch, a farm on the east side had been removed from production perhaps two years before, under a federal conservation reserve program that pays landowners who allow fields to revert to wetlands. A variety of grasses and shrubs were growing, which can filter out nitrates and other pollutants before they enter the watershed. But on the west side of the ditch the field had been planted to the very edge, and in places it was actually collapsing into the water where the stream had undercut the bank.

The question of ditching is controversial. "It's a touchy situation," says McCoy. "If these ditches weren't here, we'd probably be standing in water right now. They started draining this land in the 1700s, right on up through the twentieth century. The Pocomoke was ditched from about 1925 to 1935."

So what we have here is a common modern-day dilemma. If not for the ditches, the land would not support corn and soybeans, and if not for the corn and soybeans, there would be no poultry industry. And the poultry industry drives the economy of the Delmarva Peninsula. McCoy told me that poultry houses here produce more than six hundred million chickens a year, and despite the many grain farms on the peninsula, additional grain must be imported to provide enough feed.

Sussex County, just across the state line in Delaware, has the densest poultry population in the country. And the effects of the poultry industry reach far beyond grain farmers and chicken growers. Corporations such as Perdue, Tyson Foods, Showell Farms, and Mountaire employ thousands of local people, who do everything from butchering chickens to driving trucks. In a rural area, where decent low-skill jobs are scarce, the poultry industry has raised the standard of living of many families. And they pay millions of dollars in taxes to local governments.

But at what cost? Has the improved standard of living of poultry workers taken the jobs of oystermen, crabbers, and fin-fishermen, who can no longer make a living in the Chesapeake Bay?

Though John Page Williams, a senior naturalist with the Chesapeake Bay Foundation, believes that the poultry industry has had an impact on the health of the bay, he doesn't single it out. "The bay has suffered from a number of things: inadequate sewage systems in towns and cities along the bay, runoff from parking lots and from lawns, development along the bayfront and in the tributaries, and air pollution. They all play a role."

Williams says the extensive ditching over the years has prevented swamps and wetlands from performing one of their natural roles, that of filtering out pollutants. "In a natural system the water moves slowly, and a lot of things are filtered out by the time it makes its way through the watershed. But the system here has been changed from a filter to a funnel. The water moves swiftly through the watershed and it carries a lot of bad things with it."

Everyone agrees that getting rid of those bad things is going to be expensive. Outdated and inadequate sewage plants in towns and cities along the bay and its tributaries need to be improved or replaced. Few municipalities have the funds to do so, and fewer still the will to raise taxes and hookup fees. The poultry industry is responsible for much of the excess nitrogen and phosphorous entering the bay, but it isn't fair to expect farmers and growers to foot the bill for improvements. Residential and commercial development on and near the waterfront is responsible for non-point source pollution and destruction and damage to wetlands, yet the demand for waterfront homes and recreational amenities has never been higher.

The health of the Chesapeake Bay has been in decline for years, and although environmentalists and many politicians agree that something must be done, the prognosis is not good, and the patient still is on life support. Late-summer newspaper stories echo the same theme year after year. "Parts of bay can't sustain any life," ran a headline in the *Richmond Times-Dispatch* in August 2005. "Bay gasping for oxygen, group's readings show," stated the Norfolk *Virginian-Pilot* during the same period.

Both newspapers were reporting on studies that show the bay becomes starved for oxygen in late summer. The *Times-Dispatch* was describing the results of an EPA survey that found oxygen absent from 10 percent of the bay's main stem, a channel running from roughly Deltaville, Virginia, to Annapolis, Maryland. In addition, tests by Old Dominion University and the Maryland Department of Natural Resources found that 41 percent of the bay's main stem contained less than five milligrams of oxygen per liter of water, not enough to support rockfish, shad, and many other species of finfish. The problem was attributed to warm summer temperatures coming on the heels of ample spring rains, which flushed nitrogen and phosphorous into the watershed, fueling an algae bloom. When the water temperature rose, the algae died and used up oxygen as they decomposed.

Scott Harper, an environmental writer for the *Virginian-Pilot,* followed Bill Portlock as he took oxygen readings in late July of 2005 in the Elizabeth River, whose branches run through industrialized Norfolk, Chesapeake, Portsmouth, and Virginia Beach. Portlock, who is a water quality specialist with the Chesapeake Bay Foundation and had been watching oxygen levels decline since May, took six samples from the river, from where it flows through Chesapeake suburbs to its industrial core on the Eastern Branch in Norfolk. The best reading of the day was 5.8 milligrams of oxygen per liter, in water drawn beyond the barges and shipyards of downtown Norfolk. The worst was 2.4 milligrams per liter, from samples taken at three other locations.

Harper also reported that Virginia lawmakers estimate that it will take $2.3 billion to upgrade sewage plants and help curb farm pollution. A special committee of state lawmakers is studying possible funding ideas to foot the bill, Harper wrote.

But while lawmakers study the problem, and while conservationists wring their hands, Portlock probably had the best take on the situation. "Our challenge is to get politicians and the average Joe Citizen to realize this is not normal, this is not the way it's supposed to be," he said. "You figure at some point in time people are going to say 'We're mad as hell and we're not going to take it anymore.'"

Portlock added that the time for that had not yet come.

5
The Green Sea

WHEN WILLIAM BYRD surveyed the Virginia-Carolina line in 1729, he looked out over the area as he approached from the east and saw thousands of acres of grasses blowing in the breeze. He wrote in his journal that it looked like a vast green sea, and that term stuck. If you look at old maps, you'll see the words Great Dismal Swamp and Green Sea used to describe the vast swamp lands that once covered most of southeast Virginia and north-east North Carolina.

Today, much of Byrd's Green Sea has been converted to farmland, homesites, businesses, and highways. Take I-64 south from Virginia Beach to Suffolk, and you'll drive through it. In the median, you can still see Byrd's grasses blowing in the breeze. Now and then black gum trees will appear, their feet wet with swamp water, and you begin to realize that, yes, before the interstate, before the shopping centers and subdivisions, a swamp was here.

Great Dismal Swamp National Wildlife Refuge is the heart of what once was a much larger tract of swampland. And if you drive from Suffolk out VA 460 through Windsor, Zuni, Ivor, and Wakefield, you can get an inkling of what Byrd might have seen when he and his crew cut a line through here.

Off the highway, this still is a rural, thickly forested region drained by slow-moving rivers such as the Nottoway, the Blackwater, and, farther south, the Meherrin. The area fascinates botanists because plants grow here that are found nowhere else in Virginia, and although the area has not see the development experienced by the Hampton Roads area, it has changed markedly since the days of Byrd's exploration.

In his journal, Byrd wrote of exploring an area called Sandy Ridge near Edenton, where he found huge pine trees with long leaves and very large cones that held seeds the size of black-eyed peas. The seeds, he wrote, provided "very good mast for hogs and fattens them in a short time."

Byrd also noted that the pines had large amounts of turpentine "and consequently yield more tar than either the yellow or white pine and for the same reason make them more durable timber for building. The inhabitants hereabouts pick up knots of lightwood in abundance, which they burn into tar and then carry it to Norfolk or Nansemond for a market."

Byrd was describing the longleaf pine (*Pinus palustris*), which in the eighteenth century was common in the sandy lowlands of the Coastal Plain. Byrd likely saw thousands of acres of these stately pines, which have leaves nearly a foot long and cones much larger than those of the loblolly. The trees grew in poor, sandy soil, with none of the dense undergrowth we associate with mixed forests today. Instead, the lowlands would have seemed an almost a parklike setting, with huge trees towering over sandy soil dotted by a scattering of ferns and wildflowers.

Unfortunately, few stands of longleaf pine remain along the North Carolina–Virginia line, and few of us will ever see a pine forest as Byrd did.

But that may change if a group of conservationists and forestry experts are successful. A survey of native longleaf pine habitat in southeastern Virginia has been completed through a partnership between The Nature Conservancy (TNC), the Virginia Department of Conservation and Recreation (DCR), and the International Paper Company, which owns a large tract of pine habitat near Franklin called South Quay. The survey will help formulate a long-term management plan that will protect the pines.

"The South Quay site has been a conservation priority for years," Brian van Eerden told me. Van Eerden is a stewardship ecologist with the conservancy and has been studying the flora of southeast Virginia for some time. "The conservancy and DCR were thrilled to get the opportunity to evaluate the site," van Eerden said. "Only a few areas are left in Virginia where you can see native longleaf pine forest, and the South Quay site is the best we have. When it comes to rare habitats in North America, longleaf forests rank near the top of the list. Less than 1 percent of the original range of longleaf pine remains in all of the southeastern United States. We are fortunate to have a fragment in Virginia that survived."

The site is along the eastern side of the Blackwater River and is part of the larger Chowan Sand Ridge, a strip of land dominated by dry, sandy, rolling hills extending from Gates County in North Carolina to Isle of Wight County in Virginia. The South Quay site encompasses several thousand acres and extends southward across the North Carolina line.

"Botanists have been fascinated by this place for years," said van Eerden. "Merritt Lyndon Fernald, the famous botanist from

Harvard University, spent ten summers in the area in the 1930s and 1940s collecting and describing plant specimens. He collected orchids, carnivorous plants, and other plants that had never been seen before in Virginia. The work he did in those years was the last systematic survey. It was interesting to see how things have changed over the past seventy or so years."

John F. Townsend, who led the survey for the Virginia DCR, said the group found about 90 percent of the plant species Fernald listed. "We even found a few he didn't have," said Townsend. "The discouraging news is that one of his favorite collecting sites, which he described as a 'savannah-like swale,' is now underwater. It was flooded when Union Camp built a holding pond years ago."

Townsend added that the South Quay tract is fascinating to botanists because it has so many plants that are rare in Virginia. "South Quay represents the northern limit of many southern species," he said. "Numerous plants at South Quay are found nowhere else in the state."

Among the plants found at the tract are sandhills butterfly weed, sweet gallberry, pine-barren rush, northern sheep laurel, hairy seedbox, pixie moss, ciliate meadow beauty, honey cup, creeping blueberry, and fasciculate beakrush. South Quay is also the only known site in Virginia for the orange-bellied tiger beetle and the barrens dagger moth.

The most impressive state rarity on the site, though, is the longleaf pine. Townsend said that when European settlers arrived in the early seventeenth century, they found thousands of acres of longleaf pine forests across southeastern Virginia. The great trees did not fare well as the human presence grew. The larger specimens were used to make masts for sailing ships, and because the longleaf has such a high amount of resin, the trees were widely harvested to make waterproofing tars for the British maritime industry. In addition, feral hogs introduced by settlers destroyed seedlings, and fire suppression further disrupted the natural process of longleaf reproduction. By the mid-nineteenth century only scattered remnants of longleaf pine remained.

Botanists are now using fire as a method of restoring longleaf pines and related plants to their native habitat. Lytton John Musselman, a professor of botany at Old Dominion University, manages the Blackwater Ecologic Preserve in Isle of Wight County. Musselman has been selectively burning portions of the preserve to remove dense underbrush and provide a foothold for a wide variety of plants and young trees. Without fire, he says, the longleaf pine cannot adequately reproduce and its descendants cannot be sustained.

The tree has adapted well to harsh conditions. It grows in sandy, nutrient-poor soils, and has tough, thick bark that makes it fire resistant. Indeed, the tree depends upon fire for the nutrients and minerals that sustain life. Adult trees are covered with asbestos-like bark, and even the young buds are protected by a thick layer of needles. The shape of the tree also tends to direct fire upward, away from the buds of its seedlings. When a fire burns itself out, the ashes replenish the soil with nutrients and essential minerals. In short, fire feeds older trees, clears the way for younger ones, and reestablishes other plants whose seeds have been dormant, awaiting the right conditions.

The Blackwater preserve was established in 1985 when Union Camp Corporation donated a three-hundred-acre site to Old Dominion University. Since then, ODU has worked with the conservancy, the Virginia Division of Forestry, and the Department of Natural Resources to reestablish longleaf pines and related species through controlled burns. Since the program began, longleaf pines are regenerating, and mosses, pitcher plants, orchids, and other plants associated with longleaf habitat have reappeared. Yet others may still be dormant, banked in the soil, awaiting rebirth and baptism by fire.

Proper forest management benefits not only native trees and plants, but birds as well. In William Byrd's time, the pine forests would have been home to hundreds of red-cockaded woodpeckers, a small, black-and-white bird with a prominent white cheek patch, whose name comes from the small red tuft on the heads of the males.

Red-cockaded woodpeckers nest in cavities they build in large, living pine trees that are afflicted with heartwood disease, which renders the inner wood soft. These birds drill numerous holes around the nest cavity, and the resulting pine pitch that spreads down the trunk of the tree, which is thought to discourage predators, is a sure sign that woodpeckers are nesting in the area.

But few of these small woodpeckers nest in Virginia anymore. The bird once was plentiful in the southeastern United States, ranging from New Jersey to Virginia and the Carolinas to Florida and Texas. But the red-cockaded woodpecker depends upon old-growth pine forests for survival, and by the mid-1900s this habitat had been reduced by more than 95 percent. The woodpecker went on the endangered species list in 1970, and in Virginia fewer than two dozen individuals are believed to survive.

A few miles west of the longleaf pine preserve, near the town of Wakefield, the conservancy manages a large tract of forest called Piney Grove. Brian van Eerden says the management plan here is built around the needs of the red-cockaded woodpecker. "Red-cockaded woodpeckers have foraging and habitat needs that have been well identified," he observes. "They want a basal area of about thirty to fifty square feet per acre, and the management plan we put in place uses those numbers as goals."

Basal area, van Eerden explains, is a measurement of pine stems per acre of forest, and takes into account the number and size of trees. The reason basal area is important is that the woodpeckers prefer pine habitat of a certain type. If the forest is too thick or too thin, they won't use it.

So forest management at Piney Grove requires some serious number crunching. In developing the management plan, the conservancy found that some stands at Piney Grove are currently at the ideal basal area of 40 square feet per acre, while others have a density of as much as 120 square feet. Van eerden indicates that the thicker, younger stands of pine will be thinned, and that the plan not only calls for creating a target basal area for pines, but also recommends removing most midstory hardwoods. He notes that the basal area of hardwoods should be in the single digits, and that fire will be used to keep an open understory.

The Nature Conservancy has developed a thirty-year plan for each of the forty-eight pine stands at Piney Grove. The plan calls for a target of forty square feet of pine per acre, which provides a slight buffer in the event of hurricane, fire, or insect damage.

The forty-eight stands of loblolly and short needle pine range from about seven to two hundred acres. According to van Eerden, a stand is defined as a group of trees that has the same species composition and management history and is of the same age class.

The conservancy had red-cockaded woodpeckers in mind when it bought Piney Grove preserve from the Hancock Timber Resource Group in 1998. The 2,700-acre site has plenty of large pines, extensive forested areas, and Virginia's sole remaining population of nesting red-cockaded woodpeckers. Since purchasing the tract, the conservancy has managed the preserve for these woodpeckers and has worked with state and federal wildlife agencies to relocate birds from South Carolina to Piney Grove.

* * *

Swamps are defined by the types of plants growing in them, and as is the case with longleaf pines and red-cockaded woodpeckers, plants often support a specific community of animal life. Consider the tidal wetlands of the coast. These salt meadows are primarily grasslands, with saltmarsh cordgrass (*Spartina alterniflora*) the dominant species. Along the edges of creeks and bays, cordgrass grows in thick stands that are often up to six feet high.

On the upper marsh, where tidal flow is limited, the grass is shorter, ranging from about six inches to two or three feet, and it grows less dense, often intersperced with colonies of tubular plants called *Salicornia.* In the upper elevations, where the salt marsh joins fastland, *Spartina alterniflora* gives way to *Spartina patens,* a shorter, thicker grass known as saltmeadow hay, and *Distichlis spicata,* or salt grass.

The entire community of marshes, bays, islands, fish, shellfish, birds, and animals depends on these grasses, which form the basis of the wetland food chain by collecting the energy of the sun in photosynthesis and distributing it to myriad creatures as the vegetation dies and is broken down by bacteria. The mixture of bacteria, epiphytic algae, and the cellulose particles of digested spartinas form that nutrient-rich soup called detritus, which supports the rest of the salt marsh food chain.

The plants grow prodigiously during the summer, fed by nutrients swept in with the tides, storing the sun's energy during the long, unshaded days. When fall arrives, this chemical energy is released as the exposed stems and leaves die, the plant sinks to the marsh floor, and the bacteria attack. The small, single-celled bacteria don't ingest the spartinas, but instead digest its vegetation outside their cells, reducing the once substantial plants to progressively smaller bits and pieces.

The resulting mixture—bacteria, plant remains, larvae, free-flowing eggs, and algae stirred by tidal action into a life-sustaining broth—is consumed by protozoans that live in the shallow water, by the filter-feeding burrowing worms of the tidal flats, and by oysters, clams, mussels, nematodes, snails, insect larva, fiddler crabs, and small fish such as menhaden and mullet, which either filter the nutrients from the water or eat them with bottom mud.

Clams burrow in the sediment and send up a pair of siphons, one of which pulls nutrient-rich seawater through the digestive system, while the other expels small indigestible particles and waste. Most of the filter feeders consume detritus in this manner, drawing in the broth through hairlike cilia, through membranes, or, in the case of the marsh mussel, through a mesh of mucous threads covering the gills.

Fiddler crabs eat detritus by picking up gobs of it with their claws, sorting out the digestible particles with six specially adapted legs that cover their mouths. The tiny legs are shaped like paddles and are covered with stiff bristles that separate the large particles of food from the small. The small particles are digested, while the larger pieces are temporarily stored in a predigestive chamber, and are spit back into a claw and returned to the surface of the marsh when a sufficient amount has accumulated.

If you hike a high marsh or walk along an exposed tidal flat, you will come across several species of snails. Mud snails forage along the surface of the flat, scraping up detritus from the surface with their radulae, the rasplike teeth that pull food particles into their mouths. The marsh periwinkle feeds on the lower stems of spartinas, scraping away algae and nutrients that have collected there.

Clams, fiddler crabs, mud snails, marsh periwinkles, and the like are preyed upon by animals higher up the food chain—larger fish, blue crabs, waterfowl, wading birds, raccoons and other mammals. A clapper rail stalks the cordgrass marsh, spearing an unsuspecting periwinkle snail from a grass stem. A great blue heron waits patiently in a shallow gut, and surprises a passing killifish. An osprey circles over the open creek, dives, and comes up with a mullet in its talons. A fisherman drifts in a small boat along a tidal creek, hoping to entice a flounder with an offering of squid and minnows.

While the salt marsh food chain begins with the sun and the spartinas, the most remarkable feature of the ecosystem is that these plants are able to survival at all, much less begin a process that gives life to creatures ranging from one-celled protozoa to human beings. The salt water would literally suck the life out of less well-adapted plants, were they to find their way to the marsh. Through the process of osmosis, nature balances the concentration of particles suspended in water by moving a less concentrated solution through a membrane to a more concentrated one. If a freshwater plant suddenly found itself in salt water, the water in the plant's cells would be drawn out by the resulting osmotic pressure, and it would quickly die.

Spartinas have solved this problem by allowing a certain amount of salt to enter their cells, elevating the salt content within the plant to a slightly higher concentration than that of the surrounding seawater. In this way, the osmotic pressure is reversed; instead of fresh water moving away from the plant cells, seawater attempts to enter, inflating them and affording them strength and resiliency.

Spartinas, however, are selective with regard to the salts that they allow to enter, and screen out those that might do them harm. The most common sea salt, sodium chloride, is allowed to pass, as is a small amount of potassium, which is an important nutrient for the grass. The salts are screened by a cellulose membrane that covers the plant's roots, and excess and unneeded salts that are concentrated there are washed away by seawater and rain as it leeches into the soil. Other excess salts are secreted onto the leaves of the plant through special glands on their upper surfaces. On a dry summer day, a tiny coating of salt crystals make the spartina leaves shimmer in the sunlight. The crystals do not dry on the leaves as seawater evaporates, but instead come from within the plant as it attempts to maintain its perfect pitch of osmotic pressure.

Spartinas, like most land plants, draw moisture from the soil as water evaporates from their leaves. They breathe in carbon dioxide, ingested through cells in the leaves called stomata which open during the day when the plant is active and close at night to conserve water. Water evaporates while the stomata are open in daylight, and in so doing pulls water from the soil to replace it, much as someone might suck a soft drink through a straw. The liquid is transported through thin columns filled with spongy tissues called xylem, and the evaporative pressure placed on the water column helps give the plant strength.

Living in a saltwater environment is not the only challenge the spartinas must overcome. The dense marsh soil is very low in oxygen, and the grass must compete for this scarce element with millions of bacteria and with the higher organisms that live in the upper layers. Spartinas have solved this oxygen problem by adapting a series of air passages called intercalary canals, which transmit oxygen brought in through the stomata in the leaves down to the roots, where it is needed. If you pull up a *Spartina alterniflora* plant, you will usually see reddish mud adjacent to some of the roots, the result of oxygen reacting with iron sulfide in the soil to produce iron oxide, or rust.

So spartinas survive in a hostile environment by using the salt that would kill other plants to their advantage, creating a positive osmotic pressure that inflates the plants and strengthens them. And while water is carried upward from the roots through the xylem, spartinas move oxygen molecules downward through intercalary canals. But while this wonderfully adapted marsh grass boasts several chemical and physical mechanisms that allow it to thrive on salt water and to feed oxygen to its roots, it still must overcome another problem presented by the estuary: the daily buffeting of the tides, as well as occasional violent storms.

The physically demanding environment of the marsh has no windbreaks, so the breezes bend the slender spartina stems at will. During storms the ocean can breach the barrier islands and break directly upon the marsh, sending tons of water crashing down on the grass shafts. Even in normal circumstances, high tides flood in twice daily, bending the plants in a fast-moving current.

So spartinas must first of all be solidly anchored, something they accomplish by sending tough rhizomes through the muddy soil and interlocking root systems with other plants, finding strength not as individuals but as colonies that survive or perish together.

And the shafts of the plants must be resilient and tough, willing to bend but reluctant to break. Spartinas meets this challenge by relying on the same mechanisms they use to survive in salt water. By adjusting osmotic pressure so its cells are always fully inflated, the plant functions like a tire's inner tube, able to withstand great shock before puncturing. And the thin column of water contained in the xylem, drawn constantly through the roots by the evaporation occurring on the leaves, helps make the plant both flexible and extremely strong.

The stems are also engineered for strength. Cut one off and you'll see a tube within a tube, precisely separated by cellulose spacers. The stems carry no water or gases; their only function is to support the plant, and they run all the way from the seed head downward to the underground rhizomes.

The spartinas are a wonder of chemistry, physics, and structural engineering. While most plants would wilt within hours of being placed in the marsh environment, they thrive, and because of their remarkable adaptability, they pretty much have this particular stretch of wetland to themselves.

The types of plant that are found in a wetland are usually determined by the salinity of the water. On the coast, where salt levels can reach as high as thirty parts per thousand (PPT), the range of species is relatively limited. There are the spartinas, three varieties of *salicornia*, and black needlerush (*Juncus roemerianus*). A few inches higher in elevation, where the tides seldom reach, you'll find shrub communities of marsh elder (*Iva frutescens*) and groundsel tree (*Baccharis halimifolia*), interspersed with sea lavender (*Limonium carolinianum*), a perennial herb whose tiny blue flowers are prized for bouquets and dried arrangements.

In late summer sea oxeye (*Borrichia frutescens*) is in bloom, adding a touch of gold to an otherwise green landscape. In fall, the yellow flowers give way to brown, burr-like seed heads, and the thick, succulent leaves of the plant turn orange and red. *Salicornia* also turns various shades of yellow and red in autumn, providing a saltmarsh version of fall foliage.

As the salinity of the water drops, the variety of plants markedly increases. Big cordgrass (*Spartina cynosuroides*) is the largest of the spartinas and grows in dense stands in brackish marshes, often at the headwaters of rivers. The Pamunkey and Mattaponi have large tracts of big cordgrass, which muskrats often use to build their lodges.

Along many salt marshes of the Atlantic Coast and Chesapeake Bay are little islands called hammocks (or hummocks), where a bit more elevation supports a wider variety of plants and provides a startling exclamation within an otherwise endless meadow of shimmering grasses. There you will usually find wax myrtle (*Myrica cerifera*), an evergreen shrub with fragrant leaves and tiny bluish berries that are a favorite food of birds. (The yellow-rumped warbler is also known as the myrtle warbler because of its fondness for wax myrtle berries.) Wax myrtle is often confused with bayberry (*Myrica pennsylvanica*), which has larger leaves and is deciduous. Hammocks may also be home to red cedar (*Juniperus virginiana*), wild black cherry (*Prunus serotina*), American holly (*Ilex opaca*), and pines. Most of these small islands have been around for many years, and large ones often boast sizable stands of old pine trees, mainly loblolly (*Pinus taeda*) and Virginia pine (*Pinus virginiana*).

Hammocks have played an important role in the natural and human history of coastal wetlands. Years ago, the larger ones served as pastures for cattle, horses, and other livestock. No fences were needed, and if the tax collector couldn't make it out to count the cattle, the annual levy was lessened considerably. Red cedar is the traditional Christmas tree on the coast, and to this day a seasonal rite for many families is to cut a fragrant cedar to bring home to decorate. Halls were also traditionally decked with boughs of holly, which grows on many hammocks and in coastal forests.

We are just beginning to appreciate the importance of hammocks in sustaining wildlife. Bald eagles and ospreys nest in the old pines. Hammocks near saltwater wetlands are used as community nesting sites by herons, egrets, ibises, and other birds. In fall, these islands are important food-and-rest stops for warblers and other songbirds as they migrate north. Many of the shrubs and trees found here are seed producers, and these help migrating birds maintain fat reserves for the long flight. Hammocks also provide homes for small mammals, and these, in turn, become prey for northern harriers and other raptors that hunt in the marsh.

• • •

Pitts Creek is a modest waterway, beginning near the mouth of the Pocomoke River and meandering inland for some ten miles just south of the Maryland-Virginia line. It is an uncommonly beautiful stream, undeveloped and wild through most of its length. Although its run is rather brief, it affords us a great example of how waterways and the surrounding landscape change as salinity varies.

There is a small county boat ramp on the south bank of the Pocomoke, and if you put a canoe in there and paddle upstream, you'll find Pitts Creek about a quarter-mile ahead on the right. You'll come across spartinas here, lush wide meadows that run all the way to the pine woods in the distance. The Pocomoke enters the Chesapeake Bay a little to the west, becoming Pocomoke Sound, and the marshes in this salty environment support mainly *alterniflora*, *patens*, and black needlerush. But at the entrance to Pitts Creek the grasses are notably taller. The creek is tidal, but it drains hundreds of acres of farmland and forest, and this infusion of fresh water reduces its salinity. The grasses here include big cordgrass (*Spartina cynosuroides*), along with stands of a tall, invasive reed called *Phragmites australis*, and for roughly five miles the water flows through the wide, sweeping vista of a cordgrass swamp. In winter, muskrat lodges stand out among the rotting vegetation. Little open ponds attract black ducks and teal. Bald eagles soar overhead. As the grasses die back in winter, they open the floor of the marsh to birds such as yellowlegs, which forage among the litter of the previous season's growth. Shorebirds pass through in late winter and early spring, and these wide flats are important feeding areas during their annual migration.

A little farther upstream we find cattails and sedges, further evidence that we are leaving a saltwater environment and entering one where fresh water predominates. The bank takes on a sturdier, more well-defined edge. Grasses are replaced by shrubs and a tall plant called marsh hibiscus (*Hibiscus moscheutos*), which in summer boasts large white and pink blossoms. There are islands of floating pickerelweed (*Pontederia cordata*) and arrow arum (*Peltandra virginica*), which is also known as duck corn.

This floating vegetation, anchored by underground rhizomes, confirm that we've left the saltwater marsh. Pickerelweed and arrow arum look similar, and often grow together. Both have large, fleshy leaves, but those of pickerelweed are somewhat heart-shaped, while arrow arum's are larger and triangular. The pickerelweed leaf has many veins that closely parallel its general shape; three dominant veins and many smaller ones run through the leaf of the arrow arum. Pickerelweed has a showy spike of blue flowers that bloom in late summer.

A broad term that applies to three species of the genus *Sagittaria,* arrowhead is another water plant found in Pitts Creek, often with pickerelweed and arrow arum. Though *Sagittaria latifolia* has arrowhead-shaped leaves from which the plant gets its name, the two other species have leaves of different shapes (*S. falcata*'s are lance shaped, while *S. graminea*'s are grasslike). All three have white tripetal flowers with yellow staminate centers, and the tubers at the ends of the rhizomes are usually eaten by waterfowl, giving arrowhead its other common name of duck potato.

Perhaps the most dramatic bloom of the swamp plants belongs to the yellow pond lily, or spatterdock (*Nuphar luteum*). Common in tidal freshwater marshes, with round or heart-shaped leaves and yellow flowers that bloom in early summer, this plant emerges from underground rhizomes, with leaves and flowers that float on the water.

At about its midway point, the character of Pitts Creek changes completely. The grasses disappear and the landscape makes an incredible transformation into a southern cypress swamp. Bald cypress, cedars, gums, and pines line the shore, standing in shallow fresh water. The creek narrows, the water darkens, and the constituent wildlife changes as well, as bald eagles, black ducks, shorebirds, red-winged blackbirds, and herons give way to wood ducks and songbirds at the headwaters. Though chickadees, titmice, cardinals, and wrens live here year round, during the spring and fall the swamp becomes part of the great migration corridor, thick with warblers, tanagers, vireos, and other birds that travel twice each year between breeding grounds in the northern United States and Canada and winter homes in the tropics.

So Pitts Creek, a modest stream roughly ten miles long, offers an incredible diversity of life. At its mouth spartinas soak up energy from the sun during long summer days, and in winter they release this energy to the marsh, providing the building blocks for a remarkable variety of living things. And within a few miles it becomes a classic freshwater cypress swamp, a landscape filled with songbirds, lush plants, and trees that tower over the blackwater stream.

6

A River of Birds

IT WAS A spring evening in a remote rain forest in Panama, and something was in the air. The sense of anticipation was palpable, as if a storm was brewing, an electric, supercharged knowledge that something would be happening very soon and it would change the life of the forest.

High in the canopy the birds gathered nervously, flitting about in the half-light, calling in high-pitched song. There were warblers in brilliant gold-and-green plumage, black-throated blues, scarlet tanagers with jet black wings. Hundreds, thousands of birds gathered in the canopy, as if waiting for word to be given, waiting for the flag to drop and begin their new life.

And then a dozen took to the night sky, and another dozen, and all the others followed until, perhaps an hour later, the empty canopy fell silent. The birds were on their way north, prompted by some biological clock that tells them the time has come. Like all animals, they are driven to reproduce themselves, but with their urge to procreate comes the need to travel. They leave their winter homes in the tropical rainforests of Central and South

America, and fly en masse north across the Gulf of Mexico, across the southeastern United States, up the mid-Atlantic coast and the Appalachians, and onward to the forests of the northern United States and Canada. Researchers watching them on radar have described the mass migration as a river of birds, hundreds of thousands of warblers, tanagers, thrushes—tiny songbirds weighing only a few grams each—called each spring to the forests of North America to mate, build nests, and raise young. As summer wanes, they once again feel the need to move, and in August and September they take to the skies, tracing a similar route southward through the forests and swamplands of the coast, and finally back to the rain forests of Panama.

Most of the birds that leave the rain forest on the spring migration do so for the first time. These are young-of-the-year birds, fledged in a Canadian forest the previous summer, who defied the odds to survive the southward migration in the fall, and surrender to the mysterious genetic order that impels them to fly thousands of miles north to continue the cycle of life. Other, older birds who know the route well begin their second, third, perhaps even fourth trip, traveling a precise itinerary, moving along the same migratory route as in past years, nesting in the same patch of timber in Canada, like humans who tend to stay in familiar and comfortable accommodations while traveling.

The migration of songbirds—these huge flocks flying, usually at night—is one of the great wonders and mysteries of nature. Ornithologists have studied the phenomenon for years, but it still is not fully understood. What is the cue that sparks the mass move? How do the flocks navigate, and manage to find the same migratory passage, the same rest stops, the same nesting grounds, year after year? How do such tiny, fragile creatures survive grueling flights of thousands of miles?

These are questions that have long confounded ornithologists, and although sophisticated tracking devices such as radar have taught us more than we once knew about the movement of birds, the mysteries of migration remain. The ornithologist Frank M. Chapman, writing in *The Travels of Birds* (published by D. Appleton and Co. of New York in 1916), may have been onto something nearly a century ago when he observed that "sight may be of assistance to birds on short journeys, but, as we have seen, it would be of small service over hundreds, not to say thousands, of miles of water." Discounting their senses of smell, hearing, taste, and touch as well, he notes: "So we conclude that birds possess a sixth sense. This has been called the sense of direction. The sense of sight we know exists in the eye, and the sense of hearing in the ear, and in the nerves leading from these organs to the brain. But no one knows where the sense of direction is situated. Indeed, it is only within the last few years that naturalists have ventured to speak of sense of direction as something which actually exists."

Chapman, who was curator of ornithology at the American Museum of Natural History in New York, may have been ahead of his time. While contemporary ornithologists seldom use the term "sense of direction," most believe that birds do have some sort of internal compass that allows them to find their way, perhaps using celestial navigation coupled with sight as a cue.

In *Bird Life* (published by D. Appleton in 1914), Chapman observed that birds migrate mainly at night, in large flocks and at great heights. In Tenafly, New Jersey, in 1887, he and a friend focused a telescope on the moon during the fall migration, with the following result: "On the night of September 3 . . . a friend and myself, using a six-and-a-half-inch equatorial glass, saw no less than two hundred and sixty-two birds cross the narrow angle subtended by the limits of the moon between the hours of eight and eleven."

The full extent of this "night flight" became clear during World War II, when soldiers in the British Army began picking up strange, ethereal images they called "angels" on their radar screens. These turned out to be large flocks of migrating birds flying at dizzying heights. Half a century later, such advanced systems as Doppler weather surveillance have enabled ornithologists to study the flight patterns, densities, and movements of migrating birds in detail. But although radar has become quite sophisticated, and allows for more precise tracking, it still can't distinguish individual species. And it can't tell us how they know where they're going.

Most birds migrate to a certain extent. The northern cardinal, for example, is considered a non-migratory bird, but in winter, in the northern portions of its range, it may move southward to find food. But the real migratory birds, the ones that ornithologists see on radar and that breed in North America and spend the rest of the year in the tropics and subtropics to the south, are long-distance flyers. Roughly two hundred species—hummingbirds, warblers, swallows, orioles, tanagers, vireos, thrushes, flycatchers, sparrows, cuckoos, and nighthawks among them—have adopted this remarkable lifestyle. As a group, the long-distance migrants make up 60 to 80 percent of the birds that nest in forests across eastern North America.

And though these songbirds are dependent upon tropical rainforests for winter homes, and North American forests for summer breeding grounds, the forested wetlands of the Atlantic coast are equally vital to their survival. This wetland corridor stretches from the mangrove swamps of Florida northward to the cypress swamps of Georgia and South Carolina, through the pocosins of North Carolina and the Great Dismal Swamp of Virginia up the Delmarva Peninsula to the Chesapeake Bay marshes and the Delaware Bayshore, and on through the highlands of the Appalachians up into the Canadian provinces.

The Neotropical Migratory Songbird Coastal Corridor Study, a 1993 study of bird migration by S. Mabey and others, found that the most important corridor for migratory birds lies along the mid-Atlantic coast, from Cape May, New Jersey southward across Delaware Bay and down the Delmarva Peninsula to Cape Charles on the Chesapeake. The two capes sit at the tip of peninsulas, "migratory funnels" where birds gather to rest and refuel before crossing the respective expanses of the Delaware and Chesapeake bays.

Between these capes lie thousands of acres of swampland: the Great Cypress Swamp of Delaware, the vast Pocomoke River drainage, the Nassawango preserve, the low-lying salt marshes along the Chesapeake Bay and the forested barrier island corridor along the Atlantic coast. The *Coastal Corridor Study* found that the barrier islands and the narrow maritime forest and shrub thickets were vital to migration. Birds often are driven by storms off the coast during their night flights, and when they land in the morning they head for the first fastland—often an island thicket or pine forest—that they can find.

Dick Roberts, in his seventies and rail thin with a droopy white mustache, rode his old Fuji coaster bike around Woodland Trail at Chincoteague National Wildlife Refuge on Assateague Island one morning in mid-May, looking for birds. He'd set up a system of mist nets in a swampy area along the trail, and pedaled his bike from net to net, checking to see whether a bird had been entangled. When he found one, he worked it free, placed it in a soft white bag, and returned to the parking area, where his remote banding station was set up in the back of his Toyota truck. There he'd band the bird, examine it, record assorted vital statistics, and then ride back to where it had been captured, to set it free.

On a Friday in early May when I visited, Roberts spread his nets at 6 AM. At set intervals he hopped on his bike and rode the

1½-mile trail, stopping at each net like a fisherman checking his catch. If birds were fish, he wouldn't fare well. By 10 AM, with four hours gone, he'd snagged only five—three gray catbirds and two common yellowthroats.

He took each of these to his Toyota, where they were fitted with numbered bands, and their weights, sexes, and ages logged in a spiral binder. A few visitors stopped by to admire the birds, and Roberts talked a bit about migration, and the importance of this small patch of swampy pine woods in helping the birds move between their winter and summer abodes.

The common yellowthroats he caught are spectacular little birds, with glistening yellow breasts and throats (from which they take their name), black masks outlined in white, and olive upperparts. New feathers are needed for their long flight, Roberts explained, so migrating birds molt just before taking to the skies, with the males in the showy colors of their full breeding plumage. The catbird, its tail feathers worn and soon to be replaced, was not a long-range migrant and hadn't yet molted. A bird of the understory, where maneuverability counts for more than air speed, its wings were short and compact, its tail long.

Mid-May is the peak of the spring migration in the mid-Atlantic, and Roberts observed that five years back he would have netted dozens of birds of numerous species, not a paultry five in four hours. "It's discouraging," he went on to declare. "And it's happening all along the coast, not just here. The numbers are spiraling downward at an alarming rate. It's as though there are not enough birds left to replace the population."

He blames the population drop on habitat loss. "They're cutting and burning forests in Central America where the birds overwinter. We're losing forests in the northern United States and Canada, and here in the mid-Atlantic we're turning forests into shopping centers and subdivisions."

Roberts admits to being a pessimist, but the numbers of songbirds *have* dropped sharply in just the last few years. The population of all wildlife is cyclical, with trends that ebb and flow with time. But is he tracking just a natural movement downward, soon to be offset by an upswing in population, or is something more serious going on?

To a songbird, migration is not a process of conscious thought, but a matter of survival. It is built into the bird's life cycle, its pattern of behavior written in genetic code over millions of years of evolution. It is an automatic response to changes in the seasons, a reaction to an internal clock. It is not simply a long flight, but a defining part of the bird's life. In his book *Bird Life,* Frank M. Chapman, the ornithologist, observed that for some birds migration is part of the annual nesting ritual, no different than the urge that drives salmon and shad to leave open water and swim upriver to spawn.

Migratory birds leave their winter homes in early spring and move northward to the northeastern United States and Canada, to establish breeding territories and to find a mate. The males usually arrive about a week before the females and set up nesting sites. Most species are extremely site-faithful, often returning to the same area every year. Once a male and a female bond, they begin their first attempt at nesting in late April or early May.

The nesting period make take anywhere from seventeen to twenty-four days from laying to fledging. If a nest is destroyed early on, a mating pair will probably try again until they have raised a brood or it is too late to fledge young. Once the young fledge, they remain under the care of their parents for several weeks as they learn to fly and forage on their own. Under the best of circumstances, the young are independent in about six weeks. If a pair raises their first brood by early summer, they may attempt to produce a second. If not, they gain weight and molt their feathers to prepare for the fall trip south. Molting is important because it outfits the birds with new flight equipment. Male arrive in North America decked out in colorful breeding plumage, which will be replaced with less conspicuous colors for the return trip.

Sometime between late July and mid-August, young birds disperse and adults abandon their breeding territories as the southbound trip approaches. Migration begins sometime between late July and mid-September for most species. Bird banders in the mid-Atlantic states have found that the early migration brings the greatest variety. By late October, the predominant species is the yellow-rumped warbler, which often comprises more than half the birds captured.

In fall, with the addition of a new generation, the songbird population is considerably larger than it was in May. But inexperienced travelers and the hazards of a migratory lifestyle mean that few birds of the new generation will survive their first year.

The journey spans several thousand miles, and requires a tremendous amount of energy from creatures that only weigh between one-third and two ounces. Although there is evidence that some migrants fly nonstop from Canada to the Caribbean in a few days or less, most take several weeks. Traveling long distances at night in unpredictable weather can lead to exhaustion and starvation. Yet most migrants endure, relying on stopover and staging areas where food and shelter are readily available.

Between September and early November, the migrants reestablish themselves in their southern residences. From the vast continent of North America millions of songbirds pack into the relatively small landmass of Central America, the Caribbean, and northern South America, and though they predominate in Canada and the United States, during the nonbreeding season in the tropics they constitute only 20 to 50 percent of the total population. Despite the potential for competition, the mild, consistent climate and the variety of food provide sufficient compensation for the rigors they face as they travel south.

But the change in home territory must be a shock for juvenile birds, only six or so months old. The winter home offers not only new neighbors, but a vastly different climate and plant community than those in which they were born. As a result, the birds adapt to new conditions. The white-eyed vireo and eastern kingbird, which ate insects just a few weeks earlier, turn to a diet of fruit. The Tennessee warbler develops a taste for nectar. The migrants build up fat reserves over the winter until, in March, their dull colors give way to new breeding plumage in preparation for yet another long trip.

A week after our first encounter, I met with Dick Roberts again at Woodland Trail on Assateague Island. He was more upbeat than the last time we met, especially concerning the common yellowthroats, which are usually one of the most numerous species caught in mid-May. He had reviewed data gathered during the previous five years and happily discovered that although the total number of birds had declined, the proportion of yellowthroats remained constant, at roughly 18 percent of the total number caught and banded.

Still, Roberts had spread his nets at two sites by six in the morning, and by ten AM he hadn't caught a single bird. A group of middle-school students stopped by around eight, and he explained the banding operation and told them how it aids in tracking migratory patterns. During one final check of the nets he made his single catch of the morning, a male gray catbird which he weighed, checked for fat reserves, banded, and sent on its way.

Twenty-three nets cover the five different types of habitat that are the focus of his study, and by banding those birds he catches he hopes to discover which habitats are used by the various species that stop here at different times of the year. He gives the data he collects to the biologists at the refuge, who then pass the information on to the Patuxent Research Center in Maryland. "I have one set of nets in a forested area where they burned the understory," he says. "I have nets in a shrub thicket in the open, and others in shrubs under a forest canopy. My most productive net site is in a shrub thicket near the park service visitor center, which is just behind the beach. During the fall migration most of the birds I catch are at that site. Ninety percent of the birds flying south in the fall are young-of-the-year. They hatched that spring and have never migrated before, so they're not really zeroed in like an adult bird. Many of them get blown out over the ocean, and when they get back to land they gather in the first wooded area they come to, such as that shrub thicket behind the dunes. There are days in the fall when I'll catch twenty-five different species, mainly warblers. I get warblers there in the fall I never see in the spring. The birds I catch there have no fat reserves at all; they're depleted and exhausted. It continues to amaze me that a young bird so small can survive a trip from Central or South America up the coast and on to the northern U.S. and Canada."

Most of the birds Roberts bands are never seen again, but a small percentage are recovered, and this is how biologists learn about migration. "Sometimes another bander will capture a banded bird," says Roberts, "or a dead bird might be found with a band. Most of the banded birds I catch I banded myself. I caught a cardinal on Tuesday I banded two years ago. Last year I caught a blue grosbeak I had banded five years earlier, and it was an adult bird then. Blue grosbeaks spend the winter in Central America and return to nesting sites in the northern U.S. and Canada in the spring. This one apparently has learned a very detailed itinerary and sticks with it every year. Almost all the banded nesting birds I recapture were taken in the same area they were banded. It's a remarkable example of breeding-site fidelity. I believe some of the birds come back to the same bush year after year."

You would think that most of the birds he recaptures would be resident species such as cardinals, which stay in the area year round. This is not the case. Most are migrants, which visit long enough to rest and replace fat reserves and then continue on their way. It's a travel pattern with an almost human touch. By way of example, let's say you make a business trip to Detroit once a year. You always stay at the Radisson Edgewood on Englemont Drive, because you know how to find it, you're comfortable with the surroundings, the staff treat you well, and you've had success staying there. If, however, you lived in the Detroit area and decided to spend a night out, you might pick from dozens of hotels. And you probably would not select the same place on subsequent trips.

Dick Roberts works as a volunteer, bands through all the seasons, and operates on a small grant that he uses to buy and replace nets that deteriorate quickly over the course of the year. And he's running out of net money. "I've had two destroyed by deer this week," he says.

The Eastern Shore of Virginia is a narrow peninsula that separates the Chesapeake Bay from the Atlantic Ocean. Kiptopeke State Park sits on the bay just a few miles north of the tip of the peninsula. If you stand on the fishing pier there, you can see the Chesapeake Bay Bridge-Tunnel, which connects the Eastern Shore with Tidewater. Before the bridge-tunnel opened to traffic in 1964, ferries used to dock at Kiptopeke and carry cars, trucks, and passengers across the mouth of the bay to Little Creek. The fishing pier sits on the site of the old ferry terminal, and just offshore is a breakwater made up of seven World War II–vintage concrete troop ships, lined bow to stern, that protects the harbor from the open waters of the bay.

When Kiptopeke was a link in the regional transportation system, it supported motels, restaurants, gas stations, and other businesses that catered to travelers. One of the most prominent of these was the Tourinn Motor Court, which sat high on a hill on the south side of the road, overlooking the ferry terminal. On a fall day in 1963, a group of bird-watchers was having lunch at the motel restaurant when they looked out the window and noticed a remarkable number of birds outside. They finished eating, picked up their binoculars, and began to explore the woods, the fields, and the dune thickets behind the terminal.

They quickly realized that the birds they were seeing—warblers, tanagers, vireos, thrushes—were migrating south, and were gathering here at the tip of the Eastern Shore before making the eighteen-mile flight over open water to the southern shore of the bay. And they had an idea. What if they were to set up a banding station here in the fall, record the number of species caught, and amass information that could be used to study the movement of songbirds up and down the East Coast?

And so they did. And in just a few years we'll have half-a-century of continuous data on the fall migration of songbirds along the eastern United States. The original banding station began as a modest operation run by six volunteer birders: Fred Scott, Charlie Hacker, Mike and Dorothy Mitchell, and Walter and Doris Smith. Today, with the backing of several state and private groups, it is operated from mid-August through late November by the Coastal Virginia Wildlife Observatory, which also runs a nearby hawk observatory. Computers have helped greatly with the storage and analysis of data, and the banding procedure has been standardized to provide more reliable and accurate information.

During the early years of the station's operation, the excitement came when banded birds were recovered many miles away, and perhaps several years after they had first been caught and marked. But while such information was fascinating to birders, it didn't really provide much insight into the health, welfare, and habits of birds, other than confirming they were capable of flying great distances.

"Our goal now is to band a bird here, and catch it again two, three, or four days later," says Jethro Runco, the resident bander. "We weigh the birds as part of the banding process, and if a bird gains weight while it's here, we know it's finding food, it's foraging successfully. If it's losing weight, we know something is wrong."

I met Runco on a cloudless mid-September morning at the station, temperature in the 70s, with a bit of a southern breeze. A beautiful day to be outside, but not a good day for banding. Birds usually move about along the leading edge of a cold front, with a north wind pushing them south, but today they were holding steady, resting, feeding on berries. Runco and his three volunteer assistants checked the mist nets and came back with two birds, a red-eyed vireo and a prairie warbler.

"We've got a sexy one here," Runco shouts to the group, as he carefully removes the prairie warbler from a soft white bag, holding it by the legs and preening the feathers along its back. An adult male with brilliant yellow-and-black coloration, this tiny jewel of a bird weighs about half an ounce, or fourteen grams. Runco determines its age by checking the formation of skull plates on top of the head, blows on the belly to separate the feathers and check fat reserves, weighs the bird and attaches a band, and off it flies. An assistant records the data in a logbook.

The prairie warbler is a unique catch because it is an adult.

Usually, older birds follow the Appalachians as they travel from summer to winter homes. Juveniles follow the coast. Of all the birds caught and banded at Kiptopeke, perhaps one in twenty-five is an adult.

The red-eyed vireo, case in point, is a juvenile, hatched probably in June, and about three months old. "What we have here is a brown-eyed red-eyed vireo," observes Runco, referring to the fact that the red iris typical of adult birds is actually brown in juveniles.

Runco has twenty-one mist nets set over a few dozen acres at Kiptopeke. These are like soft, fine badminton nets running from ground level to about six feet high, supported by aluminum poles on each end. They cover a wide diversity of habitat, including forests of pine and mixed hardwood, maritime dune thickets, shrubland, and brush piles.

"Things have changed a little over the years," says Runco. "The habitat has changed somewhat and the placement of nets has changed over the years, but now we've pretty much got things standardized. We start banding on the 15th of August and stop on November 22nd. That gives us exactly one hundred banding days. That encompasses most of the fall migration here."

The banding station logs the greatest variety of birds early in the season, with mid-September usually the peak. "Right now we're seeing a lot of neotropical migrants," Runco says. "They're the long-distance travelers, going to South or Central America, the Caribbean, some to the tip of Florida. We're not seeing the temperate migrants yet, the yellow-rumped warblers and hermit thrushes; they'll come along later. On our best day so far we had twenty-six species, but I expect we'll have a few days with thirty or more. Once October comes along we'll be dominated by yellow-rumped warblers."

The old saying about the early bird getting the worm holds

true at Kiptopeke. Those that pass through early in the migration find more food than those that come later. Birds congregate in a small geographical area, and the resources are limited. Checking fat reserves is a good barometer of the birds' health. "The early migrants usually do gain weight," Runco says. "There still are a lot of insects in the area, as well as seeds and berries. So the early migrants usually are able to forage pretty well. There are millions of birds coming down the peninsula, and they tend to congregate here, so as the migration goes on, a lot of the insects and other food tend to be consumed. If we band a bird late in the season, catch it again a few days later and find that it's lost weight, we know the bird is in trouble."

Though banding results have confirmed that adult birds usually travel along the mountain ranges miles inland, while young birds follow the coast, ornithologists are unsure why this happens. Runco's theory is that birds find the traveling easier away from the water. "It could be that the birds flew along the coast the first year, had a hard time of it, and followed the mountains in succeeding years," he says. "These songbirds don't like to cross water, so it could be a response to a difficult first migration. Or it could be that their sense of direction is not fully developed. Instead of heading south they head southeast and then hit the water and follow the coast down."

Radar tracking has shown that birds fly at night, in large flocks, often at surprising heights, especially for migrants that weigh only a fraction of an ounce. They usually take off just before dark and fly until dawn. "Darkness is beneficial in several ways," Runco says. "It protects them from predators, and they use the stars to help navigate. They also use landmarks. The Chesapeake Bay would definitely be a major landmark. Birds like to fly after a cold front, flying with the wind to save energy. How high they fly depends on the weather. If it's hazy or foggy they might fly low, one to two hundred feet. But they've been tracked by radar here at 3,700 feet on clear nights, more than a half-mile high."

Even birds that are not usually regarded as strong flyers will join the migration. Runco once caught a Virginia rail in one of his nets, and several years ago a flock of more than twenty clapper rails was spotted on the bridge-tunnel. Rails are secretive birds that hide in the seaside salt marshes and are reluctant flyers at best, so to find them in open water is very unusual. "Rails don't really migrate, but there is post-breeding dispersal. They were moving, it got cloudy, and they bedded down on the rock islands of the bridge."

Other birds are built for flight. "The blackpoll warbler will leave here and fly to South America, nonstop," Runco notes.

The length of time that birds stay in the area depends on weather and foraging success. "If a bird flies in on a north wind and still has a lot of fat reserve, it probably will leave the next day, especially if the wind is still from the north," says Runco. "So it's here for twelve hours. On the other hand, we once had an ovenbird stay here for twenty days. Every time we caught it the bird had lost weight. Truthfully, I don't think that bird ever left the peninsula. It was late in the season and it was having problems finding food. It was a young bird and apparently never learned to forage well. On average, birds stay here five days, maybe a week. It depends on the weather and how successful they are putting on fat reserves."

The mortality rate of juveniles is high. "Of the birds that hatch out in early summer in the north, 70 to 80 percent will never make it back the following season," Runco observes. "Some will die on the migration south, some will die on their wintering grounds, and some will die on the subsequent spring migration. Once a bird becomes an adult, the mortality rate drops to about 30 percent a year."

The migrants face any number of obstacles. The youngsters

are inexperienced travelers and their internal compass is not always fully functioning. Some become disoriented and lost at sea. The flocks must find food to fuel the journey, and avoid obstacles such as towers, which have proliferated with the cell-phone generation. They must avoid predators, the most dangerous and prolific of which is the house cat.

Most of the obstacles that they face are the result of human activities. The massive development of open land, especially along the coast, has eliminated forests and shrub thickets where birds forage and rest during migration. Few of them can sustain themselves in housing developments and shopping malls. In the tropics, changing agricultural practices have brought about massive deforestation in the past twenty years, shrinking the area available for winter homes.

A growing problem in the east is nest parasitism by cowbirds, a native of the Great Plains that has moved east as forests have been cleared. Cowbirds remove eggs from the nests of warblers and replace them with their own, and so a blue-winged warbler might raise one chick of its own and three cowbirds. "Cowbirds are native to open spaces, but we've fragmented the forests so much they're able to penetrate the forests, which now affects species that live there," says Runco.

Our need for cheap and abundant energy is forcing us to make hard choices as we search for alternatives to fossil fuels. Wind farms are seen as a "green" alternative, but Runco says they are a threat to migrating songbirds and hawks. "They say that the propellers turn at a slow speed, but the blades are huge, and at the tip it's moving at 120 miles per hour. And the wind farms need to go where it's windy, on the ridge tops, and this is where hawks congregate too because it's where they get the updrafts. This is also the territory used by adult songbirds during migration. On the other hand, we'll never run out of wind. So it's a difficult choice. If we use nuclear power we create waste products future generations are going to have to deal with. If we use hydroelectric we're going to flood habitat by building dams. So you have to weigh the odds. What's good? What's bad?"

Runco concludes that practices we often consider good, such as fire supression, can have negative effects. "We've been suppressing natural fires for centuries, but some birds need fire to exist. The Kirtland's warbler nests only in young jack pines, and the only way you get young jack pines is to have a fire come through, burn old trees, and have new trees come in. Some birds need a clean understory, which fire provides. Some need the edge habitat that divides a burned and unburned area. It's difficult to think of fires as being good, and you hate to see these massive forest fires, but in the natural world you have many small patchwork fires and there's not the fuel available to produce these massive, destructive fires."

Three metal shipping containers, painted white, sit on the edge of a soybean field near the seaside village of Oyster, Virginia, a few miles north of Kiptopeke. A sixty-foot metal tower, looking like something from a giant Erector set, sits adjacent to the white cubicles, topped by a rectangular antenna that spins silently. I step inside the containers and enter a dimly lit James Bondish world of humming computers and electronic gizmos, their monitors aglow with undulating green and red images that slowly change form like a lava lamp.

It seems surreal, these hundreds of thousands of dollars worth of radar and electronic gear sitting in a windowless room in a soybean field, with deer casually grazing nearby. The people monitoring the computer screens are not plotting to take over the world; they're scientists tracking the movement of birds, and state-of-the-art radar is how they do it.

Mike Watson, a radar technician for NASA, wears a baseball

cap as he stares at the computer screen, his right hand deftly manipulating the mouse, clicking from image to image. It's an evening in late September (the 27th, to be exact), about an hour before sunset, and he will be here all night. Statistically, it's the peak of the fall migration, and a low pressure area over New Jersey will bring north winds, offering a comfortable ride for tens of thousands of songbirds heading south under the cover of darkness. They should show up on the screen most any time.

The door to the little trailer rattles open, and in walks burly, bearded Barry Truitt, carrying a box of Hardee's fried chicken and a diet Pepsi. Truitt, who at fifty-seven is chief conservation scientist for The Nature Conservancy, settles into a chair in front of a computer monitor, and he and Watson immediately begin the playful verbal sparring of two people who work closely together, enjoy each other's company, and share an intense common interest. "Anne had to go out tonight so she fed the dogs but didn't leave anything for me," he complains, digging into the box of chicken.

He gets little sympathy from Watson, who points out that he will be spending the night monitoring the radar in the trailer, with no one bringing him fried chicken. Truitt, mindful of visitors, motions toward an adjoining trailer that boasts a well-stocked pantry as he offers Watson a pale and flaccid French fry, which is declined.

Truitt is a friendly, outgoing, boisterous man, deeply tanned from untold hours in the field and on the boat, with a beard more salt than pepper and a voice that reverberates off the trailer's metal walls. He has spent years studying the migration of birds, once taking his setter to the Arctic to find and study nesting red knots, a sandpiper that migrates through the Eastern Shore in early spring before nesting far to the north.

Watson, slightly built and the consummate NASA technician,

seems most comfortable with his radar equipment. He offers visitors a tour of the setup, explaining what each of the humming black boxes does. They work well together—the technician and the field scientist. Truitt knows birds, and Watson the technology that allows ornithologists to track them.

Watson gives Truitt credit for creating this little outpost of radar ornithology, and while the latter's career centers more on birds and wildlife rather than the complexities of computer science, the technician concludes that the field scientist has learned a great deal about radar on his own, from evenings spent staring at the monitors, trying to determine whether a red blip is a flock of migrating birds or a tractor-trailer rig crossing the bridge-tunnel.

In spring 2003, NASA officials searching for a site for a weather radar station zeroed in on the 1,400 acres around the harbor in the village of Oyster on the Eastern Shore that were owned by The Nature Conservancy. NASA wanted to study patterns of local rainfall, and when the conservancy discovered that radar could also be used to track the movement of birds, a deal was struck. The space agency erected the radar tower and trucked in the trailers, and now, in addition to tracking weather systems, the equipment monitors the movement of birds for roughly fifty days a year, during the height of fall migration.

"My bosses were looking for a site equidistant between the NEXRAD weather station in Wakefield and the NASA facility at Wallops Island," says Watson. "This location was perfect, and in talking with Barry Truitt, when he realized radar could be used to track birds, we established a partnership."

Ornithologists have made use of radar since the early 1990s, when the nationwide NEXRAD system was used to produce the familiar Doppler scans for TV weather reports. Dr. Sidney Gauthreaux, director of South Carolina's Clemson University

Radar Ornithology Lab, was a pioneer in using weather radar to monitor bird movement. Though ornithologists were aware that songbirds migrated at night, few details were known about this spectacular natural phenomenon prior to Dr. Gauthreaux's work.

The radar system NASA proposed for Oyster, called NPOL, is the newest available, and far more powerful than NEXRAD. "When we met with the NASA scientists and they were explaining what NPOL could do, it was like the proverbial light bulb clicked on," says Truitt. "They were saying that the radar was sensitive enough to pick up individual raindrops, so I asked them about picking up birds. They said sure, it picks up birds, but we use filters to screen out birds and insects. I said, well, you can leave the filters off, can't you?"

The radar in use here cost NASA eight million dollars to develop, and it's one of the most sophisticated in the world. "This radar has the potential to advance the entire field of radar ornithology," notes Truitt. "NASA developed it to study rainfall, but the same qualities that make it good for rainfall make it ideal for birds. For example, a raindrop falls flat like a pancake—it's not a rounded drop like everyone thinks. This radar can look at it both horizontally and vertically and tell how high and wide a target is. So the potential is there to not only see flocks of birds, but individuals, and to distinguish size and shape."

A flock of birds resembles a moving storm on a TV weather report, and on this September evening, just past sunset, an avian front hit the East Coast. On Doppler radar, precipitation shows up as varying shades of green, yellow, orange, and red. Energy emitted from the radar tower travels through space, bouncing back to an eighteen-foot flat-panel antenna when it strikes airborne objects. It is then decoded by computer software, which can determine density, speed, and altitude. The greater the density, the closer the mass on the monitors is to the red area of the color spectrum.

On this particular night, a great red mass bled slowly southward from Cape May, New Jersey. And from the Eastern Shore of Virginia, near the southern tip of the peninsula, the screen began to glow, first yellow, then orange, and then crimson. Thousands of birds, perhaps millions, were up and flying, heading south under the cover of darkness, unnoticed by all but a few people huddled here in the dark, around a computer screen in a little white trailer in the corner of a soybean field.

Early the next morning I went to Kiptopeke. Predictably, Jethro Runco and his team of volunteer banders were busy. Dozens of birds were being captured in mist nets, carried in those soft white bags to the station, to be weighed, sexed, aged, banded, logged in, and released. By 10 AM the group had caught and released more than 125 birds of some twenty species.

Were these birds part of the great red mass we saw heading south from Cape May? "Some, perhaps," said Runco. "But most birds, if traveling conditions are right, will keep on moving. Blackpoll warblers will fly to the tropics nonstop. If a bird travels at forty miles per hour, and it flies for ten hours, it will cover four hundred miles. If it's near dawn and suddenly the bird faces the Chesapeake Bay, it might decide to fall out and rest and regroup. It depends upon the weather, the winds and the particular bird."

Still, it is fascinating to realize that the night before, we'd been watching many of these same birds on eight million dollars worth of radar, as they flew perhaps a half-mile high, navigating southward in the darkness. There is no doubt that in the future we will make use of ever more sophisticated systems to learn more about bird migration, perhaps even identifying individual species as they navigate through the night sky. But today, perhaps no field of research varies so greatly in its approaches and its methods of collecting and processing data. Dick Roberts, pedaling his bicycle from net to net, studying birds from the back of his Toyota truck, has provided researchers with years of data on bird migration. The banding station at Kiptopeke, operated by a long list of volunteers, is nearing the half-century mark. And here at Oyster the scientists stare at their computer monitors, analyzing glowing red forms that change shape on-screen like protozoans.

It strikes me that the value in all of this is not simply the collection of data, or driving the science forward, but comes from making a growing segment of our culture aware of the importance of birds, of their needs and habits and how their lives and ours are intertwined. If we damage their world, we also damage our own. A generation ago, bird-watchers were thought to be a bit odd, or comical, like the Jane Hathaway character on the *Beverly Hillbillies,* thin as a rail, dressed in khaki shorts and knee boots, wearing a military shirt and hat, binoculars dangling around her neck. Today, bird-watching is a mainstream sport—the fastest growing outdoor activity in America—and I believe that this is largely due to the knowledge provided by people like Dick Roberts, Jethro Runco, Mike Watson, Barry Truit, and the countless volunteers who have brought millions of people closer to the world of birds.

The last time I saw Runco he was explaining migration to a group of young people. He banded a prairie warbler, and placed it in the hand of a boy about ten years old, first showing him how to safely hold the intensely beautiful little bird, then telling him to release it. The child slowly opened his hand and for an instant the warbler sat on his open palm. And then in a golden flash it was gone. But the look in the boy's eyes was telling. Never again would he look at birds without thinking about the miracle of migration, nor would he forget that his life and theirs are deeply connected. In an instant of realization, his world and the world of the birds became as one.

7

Delicate Damsels and Dragons That Fly

I PUT IN at Porter's Crossing and was slowly floating down the Pocomoke, gradually making my way toward Snow Hill. The water is fresh here, but it is pushed and pulled by the tides, which I find fascinating. Here, in a blackwater swamp, paddling among cypress knees and water lilies, I float on a tide that links me to the Chesapeake Bay many miles away, and eventually to the Atlantic Ocean. And for that matter, to the magnetic pull of the moon.

I nudge the canoe into a water lily thicket and watch the dragonflies hunt. Here, in a world moved by ocean tides and lunar gravity, these delicate insects have a universe of their own. The females lay eggs in the quiet shallows among the lily pads. A few weeks later the eggs break open and larvae appear. The larvae, called nymphs, grow like crabs, shedding their shells with each growth spurt. Some will stay in the larval stage for four or five years, growing to two or more inches in length, feeding on the larva of smaller insects such as flies and mosquitoes, until on a warm evening the nymph heeds some mysterious call of nature and crawls up a plant stem. There, the crisp exoskeleton splits apart, and from it emerges an incredibly beautiful, athletic, fearsome winged predator: the adult dragonfly.

This to me is one of the most incredible metamorphoses in all of nature. We begin with a decidedly homely, beetlelike, waterbound insect, one most humans never see unless they're crazy enough to drag a net through the shallows to see what's in there. (All right, I confess.) It lives underwater in pond sediment for years, and if it can avoid being eaten by a bluegill or bass, it will crawl from the water, climb into the air, shed its lackluster shell, and become a bright and beautiful flying machine, one of the most skilled aviators of the swamp.

As I mentioned in chapter 1, dragonflies and damselflies seem to me to be the swamp's signature insect. And like their surroundings, they are beautiful, mysterious, complex, and little understood. Because they're colorful and graceful and pleasant to look at, they've recently become popular with the public. Not long ago, we bought a wind chime decorated with dragonflies at our local garden supply store. We have a copper dragonfly in our perennial bed that was crafted by an artist friend. We see dragonflies depicted in jewelry, on notepads, and in many items in gift shops and catalogs. Yet most people know very little about them. If you asked someone to name his or her favorite species, you would probably get a blank stare. "You mean they have names?"

This is a curious phenomenon, reflecting perhaps our general ambivalence toward swamps and wetlands and the animals and plants associated with them. We like the color and grace of dragonflies, yet we know little about how they live. Most people who aren't bird-watchers can usually name a few favorite species of birds; many of us can even rattle off the names of a few butterflies. But mention to someone that you saw a cobra clubtail this morning and see what kind of reaction you get.

So let's consider dragonflies and damselflies a little more closely. First off, how can you tell whether a particular insect is a damsel or a dragon? The easy—if sexist—way to keep this straight is to associate stereotypically feminine characteristics with damsels and masculine ones with dragons.

Damselflies usually are smaller and more slender than dragonflies. They are more delicate looking. Damselflies are not strong fliers, and they usually stay near the ground or water surface. Dragonflies are strong and fast and hunt out in the open.

The compound eyes of damselflies are separated by a distance greater than the eye diameter, while the eyes of most dragonflies meet in the middle of the head.

Damselflies, when perched, hold their wings pressed together over their back, or just partially spread. Dragonflies hold their wings flat out toward the sides. The four wings of the damselfly are roughly the same size and shape. The dragonfly's hindwings are broader than the forewings.

Look closely at a dragonfly or damselfly and you'll see a formidable predator. Imagine you're a mosquito, being chased by an insect one hundred times your size, whose head is dominated by compound eyes of some 28,000 facets that give your pursuer exceptional sight in almost any direction. Below the eyes is a huge jaw that stretches from one side of the head to the other. When the mouth is open, it can engulf you in a single deadly bite.

Above the jaws and eyes, on top of the head, are the antennae, or what remain of them. They are not long and wispy like those of other insects. The dragonfly is a highly visual creature, with a very advanced sense of sight. The antennae, which are used for touch, are not needed and over the generations have been reduced to stubs.

Behind the head, attached to the middle section (or thorax), are six legs, clustered like a basket for catching aerial prey and covered with sharp spines to keep it from escaping. Unlike other insects, the legs are located in front of the wings, just under the mouth. They are made for hunting, not walking. Dragonflies and damselflies use their legs to perch on plants, trees, rocks, and the ground, but their main function is to grab and hold prey, part of the deadly raptor's arsenal of this remarkable insect.

In addition to being outfitted with such hardware, dragonflies are outstanding fliers. They can fly forward, backward, sideways, and hover, and they are among the fastest of all insects. Most are also very colorful. For example, the twin-spotted spiketail is large and striking, with yellow stripes on the sides of the thorax and yellow spots on the dorsal (or top) portion of the abdomen. The green darner has a bright green thorax and a blue abdomen with a black dorsal stripe. The Halloween pennant is orange and brown, with boldly marked wings that make it look like a butterfly when in flight.

North America has some 435 species of dragonflies and damselflies, and learning to identify even a small number of these is a daunting task. Knowing the differences between damselflies from dragonflies is a good place to start, especially if it awakens your curiosity and encourages you to trek to a swamp or wetland and take a close look around. The more you learn about these aerial wonders, the more you realize how special they are, as are the habitats they call home.

Dragonflies and damselflies belong to the insect order Odonata, which is derived from the Latin term meaning "toothed." If you've ever looked at their mouthparts with a magnifier, you'll quickly catch on.

Perhaps no other insect has been the subject of so much superstition and folklore. Around the world, odonates are known by dozens of common names, not all of which are complimentary. In Germany, their names mean "devil's horse" or "devil's needle," "water witch," and "snake killer." In England, they are called "devil's darning needles" and "horse stingers." The Swedes sometimes refer to them as "hobgoblin flies" and "blind stingers," which stems from the belief that they can pluck out your eyes, or even sew them shut.

Not all cultures are negative toward Odonata, though. In some Asian countries, odonates are treated as holy creatures and are believed to have medicinal qualities. Dragonflies are caught and fried in oil or eaten in soups in parts of Indonesia, Africa, and South America.

Odonates include some of the oldest species of insects and were around when dinosaurs roamed the earth. Their fossilized remains are the size of hawks, with a wingspan of thirty inches. Small wonder the insect carries more than its share of superstition and folklore.

Dragonflies, or course, cannot sew one's eyes shut, nor do they sting horses and kill snakes. They have nothing to do with devils, witches, and hobgoblins. Neither are they holy and graced with medicinal qualities, as far as I know. And I won't comment on their culinary worth.

What they do have is an extraordinary life cycle, beginning as underwater predators and ending as exquisite flying machines, transforming themselves from mud-loving bugs to fast, agile, and beautiful aerial artists.

Let's begin with dragonfly sex. As with many species of birds, the males are most colorful during mating season in spring and early summer. Most males are also territorial, staking out sections of stream bank that can range from just one or two yards to close to fifty. The size of the territory usually depends upon the number of males in the area and the quality of habitat. And like birds, females are less colorful during mating season and usually hang back, avoiding open air, until they are ready to mate.

Males will jealously guard their air space. If an intruder enters, the defending male will fly at him in as threatening a manner as he can muster, a gesture that usually is more bluff than threat. The invading male will usually concede his opponent's territorial rights and take his business elsewhere.

A territorial male will, however, welcome the arrival of a female of his species. Most pairs begin mating activities with an elaborate courtship display, an activity that entomologists believe is keyed to the dramatic changes the insects undergo during their brief adult lives. Although most dragonflies are drab after emergence from the nymph stage, they gain color as they enter their reproductive period. Dragonflies and damselflies are visual animals, and their shapes and colors are keys to the reproductive cycle.

The actual mating act is unique among insects. The male and female join in the air, creating something of a flying cartwheel called a "copulation wheel." They may mate while flying, or they may alight on a reed or some emergent vegetation.

While the female's reproductive organs are rather conventional—a long, slender abdomen terminating in the sexual organs and egg-laying apparatus—the male has a more bizarre configuration. The end of his abdomen holds not just a genital opening, but also a powerful set of clasping organs. Strangely, these claspers are used to grasp the female's neck rather than her reproductive organs.

This presents a decidedly exotic dilemma. The male's genital opening is fixed to the neck of the female, but her reproductive organs are located at the tip of her tail. The solution to the problem is equally unique. Unlike any other insect, the male dragonfly has a second genital pouch beneath the front portion of the abdomen, and before mating he curves his abdomen downward and inserts sperm into this secondary sac. Then, when the male and female have linked, the female inserts the tip of her abdomen into the male's loaded sac to receive the sperm. The male dragonfly also has the unusual ability to remove sperm deposited by a previous male, thus ensuring that only he will fertilize the female's eggs.

After mating, the male stays with the female and leads her to a suitable place to deposit the fertilized eggs, usually in the water or along the lower edges of stems and leaves of aquatic plants. Females can store sperm for a long time, and the eggs are fertilized just before they are laid.

Females will lay hundreds of eggs over the course of several egg-laying sessions. Most of these are very small (less than 1 mm in diameter), and depending upon the species can be white, yellow, orange, brown, green, or gray. Although the eggs of some species overwinter, most hatch in one to eight weeks. The majority, however, do not survive to reach the larval stage, but instead become part of the detritus of the swamp, the planktonic soup eaten by larvae, small fish, and filter feeders such as mussels.

The larvae of dragonflies and damselflies are easily differentiated. Larval dragonflies are squat, thick-bodied, and colored to match their surroundings. Those living among plants are green, while the bottom dwellers are brown. Damselfly nymphs are smaller, slimmer, and have three leaflike gills at the tip of the abdomen.

Although the larvae have none of the glamour of the mature insects, they are no less remarkable. Dragonfly larvae, for example, have the ability to move through the water by jet propulsion. They have enlarged, muscular rectums and breathe by forcing water in and out through rectal gills. When threatened, they tuck in their legs and quickly force water through the gills to propel themselves to safety. Scientists have clocked some species at speeds of twenty inches per second.

Like the adults, the larvae are accomplished predators, and their taste for mosquito larvae make them welcome in areas where mosquitoes breed. The dragonfly larva is equipped with a formidable hinged jaw that can be suddenly thrust forward to grasp unwary prey. The victim is impaled, then pulled back into the sharp jaws.

The most remarkable stage of the life cycle is the last, when the insect emerges from the water, climbs into the open air, and becomes, for only a few days or weeks, one of the most accomplished flyers in the insect world. Depending upon the species, nymphs take anywhere from several weeks to five years to mature and begin the metamorphosis. When this occurs, the nymph stops breathing through the gills, climbs a stem of emergent vegetation, and takes its first breaths of fresh air, as the skin of the exoskeleton splits down the back. Gradually the adult emerges, soft and vulnerable. In time the body will harden, the wings will become firm, and the dragonfly will take to the air and begin its final, and most dramatic, stage of life.

About a year ago I decided to become more proficient at identifying dragonflies and damselflies. I'm pretty good at birds, although I do occasionally have a problem with sparrows and warblers. Warblers pass through in the spring during a window of about three weeks, when they're in breeding plumage and full of song, and by the time they've left I'm beginning to catch on. But when they return a few months later, they're silent and wearing their fall drab, and I'm at a loss. By the following spring, I have to start all over again. It's the endless "what's that warbler" cycle.

But dragonflies and damselflies are found in the mid-Atlantic area for much of the year. Adults emerge from nymphs in spring, stake out territories, breed, and are gone, to be followed by another generation. And so I can look at our odonate friends as long as the weather is warm.

For birding, I have a shelf full of field guides. Every few years a new one hits the market and I buy it. Chronologically, the progression runs from the Golden, then on to Peterson, Audubon (the one with photographs), National Geographic, Stokes, and finally Sibley. I have guides to hawks, warblers, sparrows, wading birds, and nests, as well as books about the birds of Great Britain, Mexico, and the area around Washington, DC. And then there are the videos and the audiotapes.

I also have little slipcased Chester Reed guides that belonged to my father. He was an enthusiastic birder and would jot down in these the date he first saw certain birds in the spring. His yearly first-spring sighting of a kingbird did not vary by more than a few days over a period of about two decades. My father didn't make many notes in the warbler section, an omission I fully understand.

And so I went looking for dragonfly guides to add to my groaning natural history bookshelves. Curiously, there were few to be had. I found several that were very scientific and a bit over my head, and regional guides and checklists, but I live a far piece from Kansas, Wisconsin, Bermuda, and the Algonquin Provincial Park and decided they wouldn't do me much good on the mid-Atlantic coast.

Fortunately, at around that time, Little, Brown and Company came out with *Stokes Beginner's Guide to Dragonflies and Damselflies,* compiled by Blair Nikula and Jackie Sones, with assistance from Donald and Lillian Stokes, who have produced many helpful field guides. This is not a comprehensive guide—it covers 100 of the 435 species we have in North America—but it is well organized, well illustrated, and offers a wealth of information in a small package. There are sections on basic odonate life history, and there are color-coded identification guides for three families of damselflies and seven of dragonflies. Nikula and Sones have helped me not merely to tell damsels from dragons, but to distinguish darners from spiketails.

I'm hesitant to capture these insects, even though Nikula and Sones admit that some species are nearly impossible to identify unless you have a specimen in hand to inspect with a 10-power magnifier. But there definitely is a danger in netting and handling something so fragile, and I'd hate to kill an egg-laying female just to satisfy my curiosity. And so I found a pair of binoculars that focus so closely I can literally stand up and zero in on my toes, perfect for getting up close and personal with Odonata.

Nikula and Sones recommend that beginners learn the basic shapes, sizes, colors, and habits of the ten families of dragonflies and damselflies. Later, you can worry about distinguishing various species within families. A brief description of each of the ten families follows; perhaps these will encourage you to buy a guide, learn more, and get out in the field to explore.

Damselflies

BROAD-WINGED DAMSELS. These are iridescent, medium-size damselflies with broad, colored wings that are closed over the back when the insect is perched. They are weak fliers and perch horizontally on streamside vegetation. The eyes are black or brown. The ebony jewelwing is a large damselfly with a metallic green body and broad, black wings. The American rubyspot has a dark red thorax with thin pale stripes, a bronzy green abdomen, and clear wings with a large red patch at the base.

SPREADWING DAMSELS. Spreadwings are small to medium in size and have mostly clear wings that are held partially spread when the insect is perched, obliquely. Most species have green or bronze metallic coloration on their abdomens, with a white, waxy coating on the thorax or tip of the abdomen. The swamp spreadwing is bronze/green with a pale blue spot at the tip of the abdomen, and blue eyes. The common spreadwing has a black thorax with green shoulder stripes, and a bronze abdomen with a waxy white tip.

POND DAMSELS. These are small to medium insects with clear wings. They usually are brightly colored and can be blue, green, red, orange, yellow, or purple. The eyes are colorful, with black spots. They fly low and perch horizontally or obliquely on vegetation, rocks, or the ground. The blue-ringed dancer is a handsome damselfly with a black thorax with blue stripes, and a black abdomen that has blue rings and a blue tip. The eastern forktail has green stripes on the thorax and a black abdomen with a blue tip. The eastern red damsel is a tiny, bright red insect with black markings on the abdomen.

Dragonflies

PETALTAIL FAMILY. These are large, gray-and-black or yellow-and-black dragonflies. They perch vertically on tree trunks or horizontally on the ground. The gray petaltail is very large, with a gray thorax that has pale markings on the side and dark bands on its abdomen, and is well-camouflaged when perched on tree trunks.

DARNER FAMILY. Darners are medium to large and most have clear wings, some with an amber wash. They are multicolored, with shades of brown, blue, yellow, green, and occasionally red. They perch vertically or obliquely, hanging from branches, and are strong fliers. The green darner has a bright green thorax and a blue abdomen with a black dorsal stripe. The fawn darner is brown, with pale yellow dots on the sides.

CLUBTAIL FAMILY. Clubtails have clear wings with an amber wash, and bodies that bear a camouflaged color pattern, sometimes brown or black with yellow, green, and gray. They perch horizontally on the ground, and are so named because of the flare at the end of the abdomen. The dragonhunter clubtail is black, with yellow stripes on the thorax and a thin yellow stripe on the top of the abdomen. The sanddragon is brown and yellow and is usually found along the sandy banks of ponds and streams, where it blends in. The cobra clubtail is black and yellow with a pronounced club that has yellow spots on the sides.

SPIKETAIL FAMILY. Spiketails are large dragonflies with clear wings and bold yellow stripes along a black thorax. They perch by hanging from branches and are good fliers, and their long, slender abdomens give rise to their name. The twin-spotted spiketail is very large and striking, with yellow stripes on the sides of the thorax and yellow spots on the dorsal (or top) portion of the abdomen.

CRUISER FAMILY. Cruisers are brown or black with yellow spots on the abdomen. The wings are clear. They are strong fliers and perch vertically, hanging from branches. The river cruiser is a large, handsome dragonfly with a black or brown thorax and a yellow-spotted and -striped abdomen. The stream cruiser is similar in size and shape, but with cream-colored stripes.

EMERALD FAMILY. These are medium to large dragonflies with clear wings that occasionally have brown patches or an amber wash. Most are brown or black with yellow stripes or spots and green iridescence. Males have spindle-shaped abdomens. The prince baskettail has bright green eyes, with conspicuous brown patches on the wings. The common baskettail has a brown abdomen with yellow spots on the sides.

SKIMMER FAMILY. Skimmers are the most common and easily observed group of dragonflies. Many species are very colorful, and some, such as the calico pennant and the Halloween pennant, have conspicuous wing patterns. The calico pennant has a red-and-black body with reddish-brown spots on the wings. The Halloween pennant is orange and brown, and its boldly marked wings make it resemble a butterfly when in flight. Most skimmers perch horizontally on the tips of vegetation.

8
On the Go on the Nassawango

ON THE UPPER Nassawango, the creek lies narrow and broken, interrupted by fallen trees and covered with thick undergrowth. Here the course is often undefined, disappearing into soft folds of peat, then appearing elsewhere. Deep in the swamp, where the waterway takes on the character of land and the land becomes like water, I was walking among cypress knees, looking for birds in the trees overhead, when I realized I was sinking.

The upper Nassawango, married with the land, is fit for neither foot nor boat, but a trail skirts the swamp, crosses it here and there, and makes exploration possible. On my first trip in, I thought that the water had been polluted by some unknown source. Along the banks of the stream the rich, dark soil had been stained by a greasy, orange substance that ran in thick veins all along the creekside. Some patches were light, almost pink in color. Others were a deep, angry orange that was completely out of place in a setting where the color scheme is primarily green and black.

But this part of the Nassawango is far removed from any possible source of pollution. Once outside the swamp we're in farm country, with huge fields where corn and soybeans grow, neat old clapboard houses with barns and silos and tractors stored away in sheds. The nearest town is Snow Hill, five miles as the crow flies. The county seat, its principal industry centers on judges, lawyers, and commissioners, none of whom leave orange stains in a distant wetland.

Some time later, I learned that this orange discoloration is a natural part of the swamp, and indeed has played a major role in the human history of this landscape. The streaks indicate the presence of iron, and when prospectors came across them in the late 1700s, they didn't sense pollution, but a potential for great profit.

It might well be said that those streaks of iron led to the destruction of this part of the Nassawango nearly two hundred years ago. In 1830 the Maryland Iron Company bought 4,800 acres of wet- and upland and built a huge brick furnace adjacent to the swamp, to heat the bog iron to the point where the molten iron separated from the ore. A great deal of charcoal was needed to do this—charcoal made from the cypress, pines, oaks, sweet gums, and other trees growing here.

"It looks like a mature forest now, but in the 1830s it was clear-cut, not a tree in sight," Kathy Fisher told me. Fisher is the executive director of Furnace Town Foundation, a nonprofit organization that has re-created the early-nineteenth-century village that once surrounded the huge old furnace, which alone survived. She and I walked the village paths to the church, the blacksmith's shop, the broom-maker's cottage, and the weaving house, strolling in the shade of huge oak and pine trees. Down the hill, approaching the swamp, was the furnace, and beyond that a forest of thick cypress and sweet gum. I had difficulty looking at this landscape and envisioning open spaces, homes, businesses, gardens, and offices, all permanently clouded by the acrid haze of smoldering charcoal.

And so there are three mental snapshots of the Nassawango swamp. The first is the current setting, with huge cypress and gum trees and a blackwater creek running dark but clean, a place of great natural beauty. Then there is the second image, the bustling mine town of three hundred people living in a smoky, sweaty hardscrabble environment. The third and most intriguing picture draws us back to 1829, the year before the iron company built the furnace and began cutting trees for charcoal, converting the orange sediment into valuable ingots that were loaded onto small boats and carried by the stream to places far away. What would these cypress trees have looked like then? How would the creek have flowed before it was deepened to suit the needs of commerce? What birds would be found here? What animals? And what about people? What did they do to survive here before the company store met their needs?

Kathy Fisher gets up in the morning and drives to work in 1832. In the village, the women on the staff and the volunteers wear long skirts and laced bodices over muslin shifts. As director of Furnace Town, Fisher has found her perfect job. Every day she immerses herself in the life of a working village that existed a generation before the first shots of the Civil War were fired, fixing her attention on the history the settlement, where no detail is too small to ignore.

While the original town made a dramatic change in the landscape, it existed for a comparatively short while. Fisher recounted that the bog ore was discovered in 1788 by Joseph Widener, a prospector from Philadelphia. But efforts to extract the iron were not made until the Maryland Iron Company was formed, and the state legislature offered it tax incentives to process the extracted material.

Initially, the company intended to unearth the bog iron and transport it to a furnace in New Jersey for processing, but the Nassawango site offered natural advantages that made building an on-site furnace a sound business decision. Nassawango Creek flowed through the swamp, and with the digging of a canal the iron could be shipped by boat, first to the Pocomoke River a few miles downstream, and then from there to the Chesapeake Bay and to factories all over the East Coast.

The proximity of the site to the bay also ensured an abundant supply of oyster shells, a source of calcium used in the smelting process that removed impurities from the iron ore. And perhaps the most important resource was the thousands of acres of mature trees that covered the area. Plenty of charcoal could be easily and cheaply made here, to stoke the furnace and heat the ore until it became molten metal. And so between 1830 and 1832 the furnace was built, and a town three hundred strong quickly grew up around it.

"Many of the people came from New Jersey," Fisher observed. "The area here was very much like the Pine Barrens, with very sandy soil. A furnace closed there around the same time this one was built, so they moved many of the skilled people down here. The Maryland Iron Company consisted of investors from Maryland, Pennsylvania, Maine, and New Jersey, and they had interests in several iron furnaces in the region. So it was a little like business works today—a factory shuts down and the employees are given the option of moving to a new factory in another location."

Local people did find jobs at the furnace, but many were unskilled, low-paying positions requiring hard labor. Some men cut trees and burned them to make charcoal. Others sailed skiffs to ferry loads of heavy bog ore from the swamp to the furnace, and then transported the processed pig iron down the canal in barges to waiting ships on Nassawango Creek. Among the skilled workers were those who operated the furnace, the men who knew how to maintain the temperature at the perfect superheated level, and exactly how much calcium to add to the mix of iron ore and charcoal. And there were blacksmiths, carpenters, cabinetmakers, broom-makers, weavers, printers, and other artisans, whose skills were critical to the daily life of the town and its industry.

While some historical villages tend to romanticize the life they re-create, this is not the aim of Furnace Town. "Living here had to have been a grueling experience, especially for the laborers," Fisher concludes. "There were no federal agencies to ensure worker safety. There was no health insurance, no retirement plan, and the pay was miserly. And the working conditions had to have been awful. Sulfur is released in the smelting process, and the odor of it permeated the town. And there were always smoldering fires where the charcoal was being made. So it was smoky and hot here. The furnace didn't operate during the winter months, so it was a hot-weather industry, made even hotter by the furnace and the constantly smoldering piles of timber. There was little shade, and the landscape was described at the time as a barren sand hill. Life here wasn't like something from a movie set."

Charcoal lay at the heart of the smelting process at Furnace Town. Charcoal makers, called colliers, would stack wood in a pile, then cover it with dirt and apply low, steady heat, using as little air as possible. If too much oxygen reached the wood, it would flame and burn and be reduced to worthless ash.

"The men would go into the forest, mark out a thirty-five-foot diameter circle, and begin cutting trees and building the charcoal mound," Fisher observed. "They cut everything. They were totally indiscriminate. They needed such a large volume of wood they couldn't afford to be choosy. Whatever was needed to complete the pile in the circle was cut."

Some of the wood, probably cypress, was used to build homes, stores, businesses, and other buildings. But most of the forest that covered the 4,800 acres fed the furnace. A stream was dammed just to the north, creating a pond of about 300 acres. Millraces were dug leading from the pond, one to power a gristmill where grains were processed, a second to run a sawmill where timber was converted to lumber, and a third to work the bellows that superheated the charcoal in the furnace. According to Fisher, the pond was often used by town residents to cool off during the summer, and for ice skating during the winter. Beavers were plentiful, and their hides were used to make winter clothing.

The operation seems to have flourished for several years after it opened. Lewis Walker was brought in from the Speedwell Furnace in New Jersey as ironmaster, and he took up residence in the village, in a large frame home called the Ironmaster's Mansion. By 1832, sixty people were employed to run the furnace, with others operating satellite businesses in the town. But by the middle of the decade, much of the timber surrounding the furnace had been cut, making the production of charcoal more difficult and more expensive. By then, the company faced competition from ironmakers around the Great Lakes, where better-quality ore yielded a superior final product. The owners of the company were forced to sell, and the furnace and the surrounding land were purchased by Benjamin Jones of Philadelphia in 1836. Jones also was unable to make the business profitable, and the property was sold at public auction the following year.

In 1837 the furnace ended up in the hands of Thomas Spence, who briefly gave it a new life. Spence carefully studied smelting technology, and retrofitted the works with a heat exchanger imported from England. This "hot blast" system raised the temperature inside by passing the air through a stove heated by hot waste gases as they left the furnace. This accelerated the combustion process and reduced the amount of charcoal needed, greatly increasing the overall efficiency and productivity of the furnace. It was one of the first hot-blast furnaces used in the United States.

And for a time Furnace Town prospered once more. Spence turned Walker's grand home into a boarding house, and built a new, fourteen-room mansion for his family. By then, a post office was in operation, as well as a bank and a shoemaker's shop. Spence owned a general mercantile in nearby Snow Hill, and opened a second store in Furnace Town, supplying residents with a wide variety of goods, from bricks for construction projects to cloth for making clothing.

The business succeeded for about a decade, but by 1850 Spence was forced to sell the furnace and surrounding property. Development of railroads and canals gave competitors an advantage in marketing their product, and the operation in Nassawango Swamp became unprofitable.

Spence declared bankruptcy, and the furnace was never again fired. Businesses shut down, most of the residents moved out, and over the years Furnace Town became a ghost town, a relic occasionally enjoyed by curious picnickers and picked over by weekend scavengers. Slowly, the old forest that surrounded the furnace gave way to new growth. The abandoned stores and shops and homes were lost to time, and by 1900 all that was left were a few foundations, the old slag pile, and the huge brick behemoth that started it all, curiously out of place in what by then was mature timber.

The furnace and the town that once supported three hundred

people lay forgotten, a colorful remnant of local history, remembered only in stories passed down by the children of children of people who once worked there.

There is, perhaps, fitting irony in the fact that the smoky village of 1832, with its landscape of charcoal fires and flaming brick tower, found rebirth in a forest fire. In May 1930, two brush fires destroyed the trees and undergrowth that had reclaimed Furnace Town, making the old stack clearly visible and accessible. Local residents began to realize the importance of the site to the history and culture of the area, and efforts were made to protect it. Brush and debris were cleared away, bricks were replaced, and archaeologists began investigating the old village site. The owners of the property, the Foster family, deeded twelve acres to the Worcester County Historical Society in 1961, and work began soon thereafter to re-create the village as it might have been more than a century earlier. Slowly, funds were raised, artifacts were unearthed and cataloged, and the town once again came to life.

The Furnace Town Foundation was then established as a nonprofit educational organization, and in the 1980s this group of volunteers was augmented by a paid staff. A caretaker's residence was added, and the former St. Mary's Parish House was moved to the site from Pocomoke City. Other buildings were relocated from nearby towns and farms, and were renovated and used to tell the story of Furnace Town. The old charging ramp—which allowed fillers to add ore, charcoal, and shells to the top of the furnace—was rebuilt, permitting visitors a close-up look at the hot-blast system, state of the art in 1840.

So the old village and its iron furnace are back, as are the thousands of acres of forest once clear-cut to sustain it. Just around the corner is the Paul Leifer Nature Trail, a loop of a mile or so that skirts the swamp and returns alongside the old canal where skiffs once ferried newly smelted iron to barges on Nassawango Creek. The trail—part of The Nature Conservancy's Nassawango Creek Preserve, a sanctuary of more than 9,000 acres—boasts a few huge cypress trees, along with some impressive pines and oaks, all of which date from 1850 onward. The landscape is serene and timeless, with cypress knees emerging from shallow, quiet water. It's difficult to imagine that at one time this was cleared land, that it played a role in the industrialization of America. "This forest demonstrates the ability of nature to heal itself," Fisher concludes. "Recovery is possible, but we can't think in ten-year soundbites."

9

The Great Dismal

THE GREAT DISMAL SWAMP had no great industries, and no bustling villages like Furnace Town, but has nonetheless been the setting for some of the more colorful episodes in American history since colonial times. Colonel William Byrd II explored the Great Dismal while surveying the state line dividing Virginia and North Carolina in 1728, and set down the first extensive descriptions. Byrd clearly was not impressed with the place and, borrowing phrases from the 1670 writings of the mapmaker Augustine Herrman, described it as "a very large swamp or bog . . . a horrible desert . . . [where] the foul damps ascend without ceasing, corrupt the air, and render it unfit for respiration. Never was rum, that cordial of life, found more necessary than in this dirty place."

It's interesting to note that despite Byrd's passionate dislike for the Dismal, he apparently spent very little time there. In *Byrd's Line* (an account of the colonel's survey trip published by the University of Virginia Press in 2002), Stephen Ausband notes that Byrd and other officers stayed at an inn in Edenton while the survey crew cut the line through the swamp. Byrd did not know the swamp, but was somehow convinced that the Dismal could support no life at all. "Not even a vulture would fly over it," he wrote in his journal.

I recall reading Byrd's description in my seventh-grade Virginia history textbook, which also detailed the efforts of George Washington, and later Patrick Henry, to drain the swamp. As I remember, this was one of Washington's first heroic acts, to take a worthless piece of real estate—where the "foul damps ascend"—and turn it into farmland and other valuable and useful property. Small wonder generations of twelve-year-olds grew up to be swamp-fearing, swamp-clearing adults.

Washington, of course, did not succeed in draining the Great Dismal, although his efforts are still evident. He, together with several other entrepreneurs, formed two syndicates known, respectively, as the Dismal Swamp Land Company and the Adventurers for Draining the Great Dismal Swamp. With the blessings of the Virginia General Assembly, they dug a five-mile-long canal from the western edge of the swamp to Lake Drummond, at its center.

The original purpose of the canal is unclear. Washington was a farmer and perhaps tried to grow rice in the wet, peaty soil. The syndicate owned about 40,000 acres by 1784, from which they harvested valuable timber. Several ditches have been dug over the centuries to facilitate the removal of hardwoods, and to ship out cypress shingles that were made here in the 1800s.

Another of the syndicate's goals was to create a canal linking the Elizabeth River and Chesapeake Bay in Norfolk with the Pasquotank River and Albemarle Sound in North Carolina, thereby providing an inland waterway for the shipment of goods along this north–south route. The Virginia General Assembly authorized the project in 1787, and digging began in 1793, once North Carolina followed suit. The Dismal Swamp Canal opened in

1805 and is still in use today, operated and maintained by the U.S. Army Corps of Engineers.

In spite of his ambitious goal of draining the Great Dismal, Washington nonetheless grew to appreciate its rare plants, birds, and great forests. Unlike those penned by Colonel Byrd, Washington's writings are glowing tributes to the swamp. He made at least six trips to these wetlands, and wrote that they were a "glorious paradise," abundant with fowl and other wildlife. Washington also noted that the Dismal was not "in a plain nor hollow, but a hillside." Lake Drummond, the only natural lake east of the Blue Ridge, sits about twenty feet above sea level.

No one seems to know exactly how the Dismal Swamp got its forbidding name. Some suggest that Colonel Byrd coined it—he certainly considered it a dismal landscape—but court records show that it was known as Dismal Swamp a year or more before Byrd's survey. A more likely possibility is that the description is a translation of the original Native American name.

Regardless of how the swamp came to be so known, the history and lore of the region have helped the Great Dismal earn its reputation. There are stories of runaway slaves, ghosts, pirates, and, of course, dangerous animals and venomous snakes lurking therein, and many people who live in the area seem to delight in the swamp's ill repute, perhaps even embellishing a tale or two.

On a summer weekend my son, Tom, and I decided to take the canoe up the Dismal Swamp Canal, turn west on the Feeder Ditch, and head into Lake Drummond, where the Army Corps has a campsite. We reached the lake after a paddle of about two hours and were setting up our tent when the lockkeeper came over and shared some unwanted information. "Yesterday I killed a rattlesnake big around as my arm right where you're pitching your tent."

We decided that was one snake we wouldn't have to worry about.

The corps campsite is near the shore of Lake Drummond, a shallow, remote body of water where cypress trees stand along the edge like sentinels. The lake, which sits in the center of the swamp and is the focal point of many of the legends, was named for William Drummond, the first colonial governor of North Carolina, who discovered it while on a hunting expedition that, for reasons unknown, he alone survived. Drummond served as governor from 1663 until 1667, and in 1676 was hanged, drawn, and quartered after being charged with treason for his part in Bacon's Rebellion.

With this share of baggage, I'm not sure that even today's history textbooks would dare suggest that the Great Dismal is a diverse and valuable ecosystem, with a wealth of animal and plant life. Like the lockkeeper who warned us of rattlesnakes, there is the persistent and perhaps pleasurable notion that danger lurks here. Our culture has been taught to beware such places, so much so that the term Dismal Swamp could be considered a redundancy.

Many writers over the years have visited the swamp and have tried valiantly to convince their readers that nothing is dismal about the Dismal, but few have enjoyed unqualified success. One of the most entertaining attempts was made by John Boyle O'Reilly, an Irishman whose book, *Athletics and Manly Sport,* describes a canoe trip to the Dismal in May 1888.

O'Reilly and a friend paddled two cedar canoes from Norfolk down the Dismal Swamp Canal to Lake Drummond. "There is no other sheet of water like this anywhere," he wrote, "no other so far removed from the turbulence of life, so defamed, while so beautiful."

While O'Reilly was taken with the beauty of Lake Drummond and impressed by the variety of birds he saw, he too became a victim of reptilian paranoia. He notes that while paddling the canoe he kept his pistol handy, firing now and then into the cane thickets to deter any cottonmouths that might entertain thoughts of attacking.

O'Reilly and his friend escaped the fangs of cottonmouths, but apparently fell victim to the barbs of some local pranksters, who convinced them that the cottonmouth was nothing compared to the deadly green snake, which actually is harmless. "The most deadly snake in the swamp is one of the smallest," he observed, clearly taken in. "He is about twelve inches in length, green in color, like that of the poplar tree in which he lives. We escaped him most fortunately, for before we heard of him we had deflowered many poplars of their beautiful blossoms."

I wondered, after reading O'Reilly's account, whether some of the great-grandchildren of those local pranksters might have grown up to be Army Corps lockkeepers.

Next to Florida's Everglades, the Great Dismal is probably the most celebrated swamp in American history. It's been the subject of poems, novels, and a great many nonfiction accounts. The legends and folklore seem never ending, especially for a place that really had no permanent settlement or industry, and no stable population. Or perhaps it is this very quality that adds to its mystery.

While it was home to no fixed community or industry, as was the case at Nassawango, the Great Dismal usually had a transient population. When the canal was completed in 1805, it opened up that portion of the swamp to commerce, making travel between Norfolk and Elizabeth City much easier. The spoil that was a byproduct of the digging was piled alongside, and a road of sorts gradually took shape, permitting travel by horse and by foot. As the road was improved, horse-and-buggy traffic joined waterborne vessels along the route. A fee was charged—12½ cents for horse and man, 6 cents per head of cattle—making the causeway possibly Virginia's first toll road.

Lake Drummond Hotel, also known as Half Way House, was built on the east bank of the canal in 1830, straddling the Virginia–North Carolina line. The hotel, midway between Norfolk and Elizabeth City, was literally half in Virginia and half in North Carolina, and had a rather notorious reputation which, like many of the tales about the Great Dismal, may have been embellished over the years.

Jesse Pugh, who served as superintendent of schools in Camden County, North Carolina in the 1950s, and Frank Taylor Williams, a great-grandson of the second owner of the hotel, coauthored a book about the Half Way House and the surrounding area entitled *The Hotel in the Great Dismal Swamp* in 1964. In it, they write that the hotel was noted principally for two things: instant marriages and dueling.

The legal age for marrying was lower in North Carolina than in Virginia, which made the hotel popular among young Virginia couples whose planned nuptials were perhaps frowned upon by the families. Couples would arrive by boat or by horse, the pastor would be summoned, and a brief but binding ceremony would be held—in the North Carolina wing of the hotel.

According to Pugh and Williams, the hotel was popular among duelists because of its remote location. Dueling was illegal, and participants usually chose locations that were isolated and at some distance from their homes in order to avoid the authorities. Few records exist to support the hotel's reputation for affairs of honor, but Pugh and Williams report that the anecdotal evidence is probably true. Newspapers of the day rarely published accounts of duels, unless the participants were arrested.

Though Lake Drummond Hotel was a well-known establishment in its day, Pugh and Williams don't recount what became of the building. From 1830 until 1835 it changed hands numerous times, once auctioned off by the federal government to settle a tax bill. The hotel apparently survived at least through the Civil War years. A Union soldier from Maine named Eugene Goodwin kept a diary of his travels in southeast Virginia and mentioned spending the night at Half Way House while on a march to North Carolina.

An engraving done in 1830 reveals that the hotel was a large, frame, one-story building, with a small, second-floor addition in the center and a wide front porch set a few feet from the canal. This lithograph depicts a bustling landscape with a crowded steamboat in the foreground, as well as sailboats loaded with supplies, and well-dressed ladies and gentlemen on the grounds of the hotel. A hunter and his hounds stand on the opposite bank. Only the fact that the engraving was designed and commissioned by the hotel owner might lead us to question its veracity.

Historical truths about the Great Dismal are difficult to confirm because many were passed down by word-of-mouth and oftimes romanticized in the process. The water of Lake Drummond was said to have been so pure that Commodore Perry took casks filled with it on his historic voyage to Japan, due to its ability to remain fresh and pure for months.

The lake's darker, so-called juniper water, stained the color of weak tea by tannins from the leaves of trees, was similarly deemed to have medicinal qualities. In *The Dismal Swamp and Lake Drummond—Early Recollections—Vivid Portrayal of Amusing Scenes* (published by Green, Burke, and Gregory of Norfolk, Virginia, in 1888), Robert Arnold of Suffolk touts the health-giving properties of the water. "Death from disease has never been known in that place, and it is impossible to tell what age one would attain if they were to take up their abode in it," Arnold wrote. "I have been told that instances were known where persons were found who were so old they had moss growing on their backs, and who could give no idea of their age."

Juniper water plays a role in another well-known Dismal Swamp story. A guest at the Lake Drummond Hotel was sitting at hearthside one evening when he spotted two carafes that he assumed held whiskey and water, respectively. Cutting the "whiskey" with the "water," he drank himself into oblivion, awakening the next day, with a presumably wicked hangover, to discover that he had been using clear corn whiskey to cut the amber-colored liquid drawn from Lake Drummond.

The lake is also the focal point of many Dismal Swamp legends. The Lady of the Lake is a myth made famous by the Irish poet Thomas Moore, who in 1803 penned "The Lake of the Dismal Swamp" while staying in Norfolk. The legend, which has numerous sources and variations, tells of an Indian maiden who dies just prior to her wedding. Her grieving lover believes that his lost love has risen from the grave and taken to the swamp. He follows her in and is never seen again, but is reunited with his Lady of the Lake in death:

> But oft, from the Indian hunter's camp
> This lover and maid so true
> Are seen at the hour of midnight damp
> To cross the Lake by fire-fly lamp,
> And paddle their white canoe!

Hunters, campers, and fishermen have oft claimed to have spotted the spectral craft illuminated by its fire-fly lamp, and scientists say the sightings might have some basis in fact. Some fungi in decaying wood produce a luminescence called foxfire, and on Lake Drummond, well removed from the blinding lights of more developed areas, an eerie phosphorescence is sometimes seen.

Other mythic stories involve the ancient, gnarled bald cypresses that stand in the shallow waters at the edge of the lake. One of these, whose splayed trunk might be said to resemble legs, is called the Deer Tree, since legend has it that this is a deer that transformed to elude hunters. Another version of this story instead holds that the deer turned itself into a witch who taunted hunting dogs. The witch ran into the lake and changed into a tree to avoid drowning, and the tree was unable to return to either of its previous forms.

The poet Robert Frost spent time in the Great Dismal as a young man. He had been rejected in love by a young lady from Boston by the name of Miss Elinor White and, heartbroken, traveled to the Great Dismal with the intent of ending his life in Lake Drummond. Instead of joining the ghostly white canoe and its fire-fly lamp, the poet turned and took the canal towpath to Kitty Hawk and Hatteras. Apparently the situation wasn't as dire as he thought. Miss Elinor White soon after became Mrs. Robert Frost.

Slavery also played a major role in the history and legends of the Great Dismal. Some accounts assert that just prior to the Civil War there were as many as a thousand slaves taking refuge in the swamp, some working with shingle crews, who brought them supplies. Dogs were specially trained to hunt down runaway slaves, something that Henry Wadsworth Longfellow recounts in his 1842 poem, "The Slave in the Dismal Swamp:"

In dark fens of the Dismal Swamp
The hunted Negro lay
He saw the fire of the midnight camp,

And heard at times a horse's tramp
And a bloodhound's distant bay.

In 1839, Harriet Beecher Stowe wrote a novel about this same situation, basing her book on the story of Nat Turner, who lived in nearby Southampton County. Like *Uncle Tom's Cabin,* this lesser known work, entitled *Dred: A Tale of the Great Dismal Swamp,* gained sympathy for the abolitionists' cause.

Robert Arnold, of Lake Drummond fame, lived during the time of slavery and includes several stories of runaway slaves in his book. He tells of a slave who fled from his owner, Augustus Holly of Bertie County, North Carolina, prompting Holly to offer a $1,000 reward for his capture. The slave was found hiding in the Great Dismal. "I saw the negro when he was brought to Suffolk and lodged in jail," wrote Arnold. "He had been shot at several times but was little hurt. He had on a coat that was impervious to shot, it being thickly wadded with turkey feathers. Small shot were the only kind used to shoot runaway slaves."

Alexander Hunter, a writer and sportsman from Arlington, Virginia, visited Lake Drummond and documented his experiences in *The Huntsman in the South* (published by Neale in 1908), an anthology of stories that he had written for newspapers and sporting magazines. The narrative about the Great Dismal is undated, but Hunter most likely visited in the 1880s; he died in 1914 at age 73, and appears to have been a middle-aged man when the account was written.

Hunter and a companion stayed in Suffolk and traveled into the swamp from the western boundary, probably via Washington's Ditch. They made their way in a small skiff paddled by a local guide identified only as Bob. Hunter notes that the waterway was tight, with a narrow footpath alongside. The Dismal Swamp Canal was plied by longboats carrying shingles made in the inner reaches to the railroad in Suffolk. Hunter wrote that the boats "were propelled by men walking on the bank and pushing a pole, one end of which was fastened to the boat. The towpath consisted of but a single log laid down, the butt of one touching another. These logs were not fastened, but lay loose in the ooze of the swamp; and although the boatman had the oar of the lighter (the longboat) with which to steady himself, he stumbled every other step, and walked in water over his knees."

According to Hunter, the cutting of shingles was the swamp's main industry. Workers would go into the Great Dismal, cut the cypress trees, and split and cure them before sending them to market: "Shingles made from (cypress) are the best and most durable of all; the wood splits readily, is soft when green, and hardens as it dries." Hunter observes that all the shingle makers were black and lived in "comfortable shanties on the little islands." The foremen were white and at workday's end returned to their homes outside.

Hunter was a bit of a dandy, and his writing often demonstrates an acerbic wit, especially when it comes to local people who were not up to the standards and manners of his social position. His description of an extended family of very poor people living in the Dismal is extremely unkind, though Hunter was probably trying to be funny when concluding that among the men, women, children, dogs, cats, and pigs living in the house, the pigs were the cleanest.

He was nonetheless clearly taken with the landscape of the Great Dismal, and goes on to offer this endorsement: "Amid the varied and vast wonders of Nature's handiwork none is more worthy the traveler's attention, the artist's pencil, or the interest of the capitalist than the Great Dismal Swamp of Virginia. No one, unless he has visited the spot, can form an adequate idea of its attractions; it must be seen to be appreciated; no description, however faithful the portrayal, no pencil, however true its limning, can do it justice."

10

A Piece of Land

AND SO MY real estate friend Carole Gubb called. "Boy, have I got a deal for you," she said.

And she did.

"You found my fifty acres of worthless swampland?" I asked.

"Not quite," she replied. "It's not fifty acres and it's not worthless."

Carole had found a wooded parcel of a little more than three acres that had been part of a farm when it was subdivided some twenty years ago. It was on a small cove on Pungoteague Creek and had a nice salt marsh with black needlerush and cordgrass. Wax myrtles and cedars grew on the higher elevations, and beyond that was a mixed forest of pine, oak, dogwood, sweet gum, holly, swamp magnolia, and maple. The understory had small holly and gum trees and wild blueberries.

And there was a swamp. No Great Dismal, mind you, no Nassawango. This was more of an accidental swamp. When the farm was originally subdivided, a pond was dug on the northeast corner of the property to hold diverted storm water. The rain still gathers here, nutrients and sediments settle out, and when the water rises to a certain level it passes through a dike, enters a swampy stretch, and slowly makes its way to Pungoteague Creek. So, in a short distance, the stream changes character from fresh water (with ferns and cattails) to salt water (full of spartinas and needlerush).

The pond, of course, is not a natural feature of the property. It was created for a purpose, and was not intended to add to the natural qualities of the place, or even to enhance its aesthetic value. But after twenty years of Mother Nature doing her duty, it is integral to both the natural qualities of the land and the aesthetics as well.

"Only three acres, and in a subdivision?" I asked Carole.

"Why don't you take a look," she answered.

And so Lynn and Tom and I walked the parcel on a hot July afternoon. With the vegetation at its leafy peak, it was difficult to get an idea of the lay of the land, but the little pond was patrolled by dozens of dragonflies, and we could see sunfish just under the surface. A green heron flushed from a wax myrtle thicket, and we heard a yellow-billed cuckoo back in the woods. The place had possibilities.

The farm, known locally as Red Bank, once belonged to the late Floyd Nock, a farmer, architect, historian, and conservationist. The term "subdivision" has a negative connotation for many of us, conjuring images of suburban sprawl, of soybean fields converted to cookie-cutter lots fanning out around cul-de-sacs. But Floyd was a bit of a visionary who had a feel for the land. Rather than calling in engineers and surveyors to determine the boundaries of each parcel, he let the land itself do the work. The three acres we looked at were bordered by a small stream on one side, and on another by a little finger of land that jutted into the creek. Floyd created about twenty parcels in this fashion, none the same shape, each with its own character.

After numerous visits by foot and by boat, it became clear that this was not just a place for weekend walks; we wanted to live here, and have our morning coffee as we watched dragonflies police the quiet waters of the pond. We made an offer to Anne Nock, Floyd's widow, and on the early October morning of our twenty-fifth wedding anniversary Lynn and I became the owners—or more accurately the current stewards—of parcel fourteen at Red Bank.

Owning land brings with it a surprising level of responsibility, even if the land is of modest dimensions. What we decide to do with our land affects not only our three acres, but the neighboring parcels as well, and the waters of Pungoteague Creek, which joins the Chesapeake Bay a few miles to the west. Our decisions affect the pond and the life it supports. If we thin the woodland and let more sunlight in, we will change the species of plants that grow there. How, then, to enjoy our land, to use it and make it part of our daily lives, without causing harm?

Increasingly, local, state, and federal governments are passing laws to help protect water quality. When we decided to build a home on our parcel, we had to place it at least one hundred feet from the nearest tidal marsh or waterway. The pond, because it was man-made, had no such requirement. Before removing trees to create a house site we had to pay $100 for a disturbance permit, and we had to determine soil types, vegetation, patterns of storm-water runoff, and various other factors that might affect the waters of Pungoteague Creek. We had to make a map of the property, showing the one-hundred-foot setback, the site to be cleared, any existing roadways and drainage ditches, and the placement of the well and septic system, and we had to indicate where we would locate the silt fencing, a plastic barrier that traps sediments that might be washed streamward during heavy rains.

I didn't mind doing all that, didn't even begrudge Accomack County the hundred bucks, and I'm glad someone in government is finally paying attention and taking water quality seriously. I suspect, though, that these concerns are unevenly addressed. I say this because over the previous year I'd spent a lot of time in northeastern Maryland and in Delaware and saw mounds of chicken manure piled high along deep ditches that empty into the Pocomoke River and the Chesapeake Bay. These locales required no silt fencing to prevent manure from washing into the waterway, and there were no holding ponds to trap sediment. There were no grassy buffer strips to protect the waterway, nothing to prevent all that waste from flowing unimpeded into the watershed. I don't mind abiding by local, state, and federal regulations if it means protecting the Chesapeake Bay, but I came away with the feeling of being bullied. I have no political clout, no political action committees, and I can't afford to donate money to politicians. I had the feeling of being an easy and convenient target, the ninety-six-pound weakling confronted by the schoolyard bully.

I call it axe-scaping, landscaping with an axe, or more frequently a chainsaw. In the thick woods, overgrown with spindly loblolly pines and sassafras, we begin the process of elimination, giving growing space to trees that seem stronger, more resilient. Timber was cut here a generation ago, and the mixed forest that replaced it was never thinned or otherwise managed, and so we have little pines, each with a three-inch-diameter base, thirty feet tall, topped by a modest plume of pine needles. Many of these trees had so few needles that they died because they could not adequately photosynthesize. Like the pines, the sassafras trees are also unnaturally tall and spindly, capped instead with just a cluster of leaves. So we remove the dead and the dying and soon approach the margin of Pungoteague Creek, opening up the view of our little cove, the woodland edge lined with saltwater bush and wax myrtle, which we keep because its berries are a favorite of yellow-rumped warblers.

Cutting trees and dragging brush is hard, satisfying work. I take down a few trees, cut the trunks into manageable sections, remove the tops, and then turn off the saw and begin dragging the pieces. In no time my T-shirt is soaked with sweat, my jeans are streaked with sticky pine sap, and I settle into a rhythm of cutting, dragging, and stacking, punctuated by water-bottle breaks. A black snake gives me the eye, its sleek body meandering around a highbush blueberry. I return to my work, and when I look back the snake is gone.

As I cut and drag, cut and drag, the woods open up and I can see where the land falls away into a low, swampy area, dropping perhaps ten feet in elevation. To the north, I can see the creek, perhaps a quarter-mile wide, quiet and serene.

I get satisfaction from opening up this woodland, making the landscape readable. And I get satisfaction from the work and sweat. When Lynn and I were first married, we built a house on Warehouse Prong, a little marshy stream at the headwaters of Pungoteague Creek, not far from here. There we went through the same process, cut and drag, cut and drag. I was in my thirties then, and now I'm sixty, so I enjoy knowing that I can still clear the brush. I ruptured a tendon in my knee six months ago and had surgery, and for three months I was immobile. When I finally took the brace off, my quadricep muscles were atrophied. In the shower my leg jiggled like a woman's breast. And so this is an act of renewal, perhaps even defiance. Cut and drag, cut and drag.

Before we set hand to chainsaw, we made a careful plant inventory of the property, noting trees that should be saved, and those that should go. We found a cluster of pink lady slippers, wild orchids that bloom in May and add to the beauty of the woods. These we carefully delineated. When we finish eliminating the weak and the dead, we'll begin the process of addition, of native plants, seed producers to attract wildlife, shade-loving plants of all sorts. But for now it is a time of removal, landscaping by elimination.

After clearing and opening up the forest somewhat, we decided where we wanted the house to be situated, on high ground overlooking the cove, with a path leading down to the pond. We solicited bids for clearing the house lot, and found that most of the contractors intended to cut down the trees, remove the stumps, and truck the waste to the local landfill, an expensive proposition in that the county charges by the pound to dispose of debris. One contractor proposed to burn this on-site. Wilson Custis suggested selling the pines to a small lumbermill nearby and marketing the hardwood for firewood. He won the bid and saved us thousands of dollars, and the trees became lumber or heated someone's home. Taking this money-saving approach to land clearing one step further, we set ourselves a challenge—to clear the house site, thin the woodland, and remove unwanted debris without making a trip to the landfill or starting a fire.

We began by laying out a system of short walking trails, starting at the rear of the house and going down to the pond, running from there along the swampy stream to the edge of Pungoteague Creek, up the northern property line, and completing the cycle by returning to the house. The spindly, arrow-straight little pines, with their tops removed, made perfect trail liners. We put them side by side, about three feet apart, and snaked the path through the woods and along the creek shore. We built a small wooden platform on the edge of the pond, added a bench, and that became the morning coffee and observation post.

But what to do with the pine tops and the other vegetation we had to remove? Carl Gray, an enterprising young man who runs a company called Best Ever Lawn Care, came up with a solution. Carl had just bought a huge used chipper from a tree-service company, and in an afternoon he and his chipper converted four massive piles of brush into two mounds of mulch, which we spread along the walking trails linking the house, the creek, and the little pond with its community of dragonflies and bullfrogs.

Bullfrogs squat Buddha-like in the shallow water of the pond, head and shoulders above the surface. If I approach the water slowly and quietly I can get close enough to get a good look at them. They are dark green, stocky, with sullen eyes, and they sit in the green underwater grass and wait for prey to pass. Our pond is thick with them; a half-dozen populate this shallow corner, and in deeper water tadpoles dimple the surface when they come up.

If I come upon the pond too quickly, the frogs will squawk loudly and dart across the water in a mad dash. They always startle me when they do that, even though I know they're there. I enjoy having the bullfrogs around, but I may be in a minority. In some quarters, bullfrogs are seen as an invasive species, like the giant reed *phragmites* is here on the East Coast.

Bullfrogs are native to the east, but not to western states. They were introduced in the west during the Great Depression as a food source, sort of an aquaculture project for amphibians. The frog ranches met with little economic success, but the bullfrog population took off, and the prolific species is blamed for a drastic decline in the number of native amphibians and reptiles in the Northwest.

The bullfrog is the largest true frog in North America. It can measure eight inches in length and leap up to three feet. It has a prodigious appetite and thrives in man-made reservoirs, storm-water ponds, irrigation ditches, and natural streams and swamps. It eats most any animal smaller than itself: insects, frogs, tadpoles, fish, small snakes, turtle hatchlings, newts, salamanders, bats, hummingbirds, and ducklings. Though all bullfrogs use their sticky tongues to subdue insects and other small prey, the large ones are more likely to lunge at their target, grasp it in their wide jaws, and use their front feet to shove the victim down the gullet.

The Washington State Department of Fish and Wildlife advocates active citizen involvement in controlling bullfrogs. On its Web site, the department offers detailed instructions on how to dispatch them: "Shooting adult bullfrogs using a single-shot .410 shotgun has been successful. Alternatively, a pellet gun or a bolt-action .22-caliber rim-fire rifle and dust-shot bullets can be used at close range. Use only dust shot in the .22, not conventional ammunition. Because all shot has the potential to ricochet, exercise extreme caution when discharging firearms on or near water."

The Web site also points out that a license is not required for hunting bullfrogs, there is no bag limit, and the season is open year round. Perhaps they should also include recipes.

Because bullfrogs are native where I live, I think I'll let them be and enjoy that booming mating call on late summer evenings.

In addition to its resident population of bullfrogs and tadpoles, the pond is home to a huge mud turtle, many small sunfish and bass, at least one muskrat, and countless dragonflies, and the myrtle thickets are filled with birds. Cardinals nest here in the summer, and in winter yellow-rumped warblers feed on the purple berries of wax myrtles. Deer appear in the early morning to drink water. We scattered some corn in the shallow end and were rewarded with regular visits from wood ducks, and we built a nesting box for them. On the creek, as the weather turns cooler in the fall, there are Canada geese, ring-necks, pied-bill grebes, and bufflehead and ruddy and a few black ducks, the latter of which keep their distance.

I wondered what effect thinning the pine woods would have, and discovered that it seems to have benefited the songbird population. The birds were numerous a few days after I hand-thinned the woodland; it was mid-October, and during a brief census one afternoon I counted titmice, Carolina chickadees, a downy woodpecker, pine warblers, and a blue jay. When we thinned trees we created a brush pile in an open area, and within a matter of minutes the white-throated sparrows were exploring it.

While we enjoy the bullfrogs, dragonflies, and the numerous birds, the creek has become the focal point of our recreation and exploration. The little pond is a self-contained community, a world within a world. Here frog eggs hatch and produce tadpoles, which after a year or so grow to full-fledged adults. They patrol the same area in which they were conceived. Dragonflies deposit their eggs on the surface of the pond, and these become larvae, which will live in the pond for years, until for a few months they become adults and produce another generation for the neighborhood.

On the creek, though, if I put my canoe in and head east, I can paddle up to the headwaters, where many years ago there was a gristmill, powered by the slow-moving water. Or I can head up Warehouse Prong, that branch of Pungoteague Creek named for a tobacco warehouse once situated near the banks. The prong is a beautiful place to paddle, still a bit wild and meandering, and in late summer acres of pickerelweed will be in bloom. In winter, when the vegetation dies back, I can glide along nearly to VA 178. What remains of Warehouse Prong passes through a culvert in the road here, to join a pond that local farmers use for irrigation.

If I leave home and head west, the possibilities are limitless. Boggs Wharf passes on my left, Evans Wharf on my right, and then I clear the little village of Harborton, where kids will be fishing for crabs off the old boat dock. After I pass Harborton the creek widens and becomes a significant body of water. Years ago, before there were paved roads, and before the New York, Philadelphia & Norfolk Railroad laid track, this was the avenue of commerce. The place names still reflect this vanished history. There no longer is a warehouse on Warehouse Prong. There are no wharves at Boggs Wharf and Evans Wharf, save for a few private boat docks. Warehouse Point, at the very mouth of the creek, has no warehouse, though a U.S. Coast Survey sketch done in 1853 shows a steamboat wharf on a little finger of land north of Buckland Gut called Warehouse Point, and with a little imagination I can picture a place where farmers would bring crops for shipment. There would be large docks with mule teams loading potatoes and tobacco onto ships, and offloading furniture, bolts of cloth, and other commodities from other ports.

Beyond Warehouse Point, Pungoteague Creek flows past Klondike Point and Bluff Point and becomes part of the Chesapeake Bay. The water here has seen New York. It has tumbled down mountain streams and flowed through languid rivers on its way to this very spot. And from here it will flow north and south, pushed by tides into the bay to Annapolis, up the Patapsco River to Baltimore, up the Susquehanna through Pennsylvania and back to New York. It will flow through the Chesapeake & Delaware Canal near Chesapeake City to the Delaware River south of Wilmington and Philadelphia.

As the tide turns, the water will flow out of the bay, to be joined by water flowing from the Potomac, the Rappahannock, the York, and the James. This great outflowing tide will cross over and under the Chesapeake Bay Bridge-Tunnel, and just as Pungoteague Creek joins the bay between Klondike and Bluff Points, the Chesapeake Bay will join the Atlantic Ocean between Cape Charles on the north and Cape Henry on the south.

From here it all falls away beyond the horizon, going where only imagination and faith can determine. It is said that people who live along the coast have great faith, and perhaps that is so. Implicit in our choosing to live here is our acceptance that there is a higher power, an invisible hand that ensures the tide will rise for only six hours, and then ebb. And though many of us use different terms to define and describe that power, we know that something beyond us controls the rise and fall, the ebb and flood.

On a warm spring afternoon on Pungoteague Creek, I take off my shirt and lay back in the canoe, resting on the gunwale. I feel the warm promise of the sun, the persistent cold bite of the breeze. Here on these quiet waters, so remote and removed from the rest of the world, I sense the pull of the tide, the power of ebb and flood, the force that holds our universe together.